Building Social Equity with AI

Validating User Transactions with AI

Raghu Banda

Building Social Equity with AI: Validating User Transactions with AI

Raghu Banda
Union City, CA, United States

ISBN-13 (pbk): 979-8-8688-0090-0   ISBN-13 (electronic): 979-8-8688-0091-7
https://doi.org/10.1007/979-8-8688-0091-7

Copyright © 2024 by Raghu Banda

Managing Director, Apress Media LLC: Welmoed Spahr
Acquisitions Editor: Shivangi Ramachandran
Development Editor: James Markham
Coordinating Editor: Shaul Elson

Cover designed by eStudioCalamar

Distributed to the book trade worldwide by Apress Media, LLC, 1 New York Plaza, New York, NY 10004, U.S.A. Phone 1-800-SPRINGER, fax (201) 348-4505, e-mail orders-ny@springer-sbm.com, or visit www.springeronline.com. Apress Media, LLC is a California LLC and the sole member (owner) is Springer Science + Business Media Finance Inc (SSBM Finance Inc). SSBM Finance Inc is a **Delaware** corporation.

For information on translations, please e-mail booktranslations@springernature.com; for reprint, paperback, or audio rights, please e-mail bookpermissions@springernature.com.

Apress titles may be purchased in bulk for academic, corporate, or promotional use. eBook versions and licenses are also available for most titles. For more information, reference our Print and eBook Bulk Sales web page at http://www.apress.com/bulk-sales.

Any source code or other supplementary material referenced by the author in this book is available to readers on GitHub (https://github.com/Apress). For more detailed information, please visit https://www.apress.com/gp/services/source-code.

If disposing of this product, please recycle the paper

To my family, whose unwavering support and encouragement have been my guiding light; to my mentors and colleagues, who have inspired me to pursue excellence and innovation; and to every individual striving for a fairer, more equitable world – this book is for you.

In loving memory of my father, Gopal Banda, who passed away in 2018. Your resilience and lessons about never giving up in life continue to inspire me every day. This work is a testament to the values you instilled in me.

To my mother, Veda Banda, your love and strength have been a constant source of comfort and motivation. This book would not have been possible without your enduring faith in me.

Table of Contents

About the Author

Raghu Banda is a senior director of AI product strategy and an enterprise architect advisor at SAP Labs, where he is responsible for digital transformations with enterprise customers. He began his career as a software developer and architect in India before moving to the United States in 1997. He joined SAP in 2001 and worked in various roles such as engineering development, customer support and implementations, and product marketing and product management. He has worked with predictive analytics and machine learning since SAP entered this arena in 2013. He holds a Bachelor of Science in computer science and engineering and recently graduated from the prestigious INSEAD business school in global leadership and international management. He is also an AI/GenAI strategist, a panelist, a regular speaker, and an advisor in the AI space.

He coauthored a book, *Implementing Machine Learning with SAP S/4HANA*, during the COVID-19 pandemic and runs a podcast "XTrawAI: Machine Learning and AI Applications" interacting with various guests on the topic of AI and ML.

About the Technical Reviewer

 **Raj Nukala** is a seasoned technology professional with over two decades of experience in engineering management and innovative solution development. He has a unique talent for blending technological expertise with strategic business acumen, which has led to the creation of high-impact teams and successful, revenue-generating products.

Raj has worked across diverse industries, including consumer and industrial markets, healthcare, telecom, and utilities. His current role as interim CTO/advisor at Bora Inc. showcases his ability to lead groundbreaking projects. At Bora, he spearheaded the development of a pioneering self-service beach chair rental system, integrating cutting-edge technology to create a seamless user experience.

In addition to his role at Bora, Raj is a senior smart grid solutions architect manager at Itron Inc., where he manages a team of solution architects. He supports multiple-utility customer smart grid initiatives across the United States and Canada, working closely with major utility companies to deploy smart grid and IoT technologies.

Raj's career also includes significant roles at IBM/Lotus and KPMG and as cofounder of NetPrise Inc., where he launched an on-demand integration platform for the healthcare industry.

Raj holds an executive MBA from Quantic School of Business and an MS in computer science from Widener University. His thought leadership is reflected in his speaking engagements at industry forums and seminars, where he shares his insights on IoT and smart grid technologies.

Raj lives in Macungie, Pennsylvania, with his wife, children, and two dogs. His story is one of relentless innovation and leadership, making him a notable figure in the technology sector.

Acknowledgments

Writing this book has been an incredible journey, and I am deeply grateful to the many individuals and organizations who have contributed to its creation.

First and foremost, I want to thank my family for their endless support, patience, and love. To my loving wife, Bindu, and my children, Tavishi and Vidushi, your belief in me has been the foundation of my endeavors. Your encouragement and understanding have been my greatest strengths.

To my mentors and colleagues, your insights and guidance have been invaluable. Special thanks to the brilliant minds and beautiful skills of people around me wherever I go. Your dedication to pushing boundaries and exploring new horizons has been a constant source of inspiration.

A heartfelt thank you to the editorial team at Apress for their professional support and commitment to excellence. Your expertise and attention to detail have elevated this work to new heights.

I am deeply appreciative of Raj, a dear friend, for his detailed review and invaluable feedback. Your thorough and thoughtful insights have significantly enriched this book.

Finally, to my readers – thank you for embarking on this journey with me. Your curiosity and passion for creating a more equitable world drive the purpose of this book. May it inspire you to innovate, advocate, and transform the systems that shape our lives.

Introduction

In an era where technology permeates every aspect of our lives, the quest for social equity has never been more critical. The intersection of artificial intelligence (AI), blockchain, and advanced data analytics offers unprecedented opportunities to address long-standing disparities and create systems that are both fair and efficient.

This book, *Building Social Equity with AI: Validating User Transactions with AI*, delves into the sophisticated processes and pioneering technologies that underpin the development of a Social Equity Score (SES). The SES is more than a mere numerical value; it is a comprehensive measure of an individual's or organization's societal contributions, encompassing factors such as environmental impact, social responsibility, and ethical practices.

Through compelling real-world examples and detailed case studies, this book explores the transformative potential of AI-driven language models, blockchain technology, and Generative AI (GenAI) in fostering social equity. From the bustling neighborhoods of São Paulo and the vibrant streets of Rio de Janeiro to the diverse communities of Mumbai, New York, London, Lagos, and Cape Town, the narratives of individuals like Diego, Maria, Priya, John, Fatima, and Amina illustrate how these technologies can build more equitable and supportive communities.

As you journey through the chapters, you will uncover the "secret sauce" behind the SES, the metrics and measurements that define it, and the innovative applications that bring it to life. You will gain insights into practical implementations, from developing personalized educational materials to creating AI-enhanced matchmaking tools for professional networking.

The concluding chapters envision a future where large language models and AI continually adapt and refine our understanding of social equity. By leveraging these technologies, we can construct systems that are not only intelligent but also just and inclusive.

Whether you are a scholar, a practitioner, a policymaker, or simply passionate about creating a fairer world, this book offers a comprehensive guide to understanding and implementing the Social Equity Score. Together, we can harness the power of technology to build a society that values quality, merit, and the intrinsic worth of every individual, every entity, and every organization.

CHAPTER 1

Social Equity and Its Significance

In today's fast-paced and ever-evolving world, the concept of social equity has become increasingly important. Social equity, in essence, means ensuring that all individuals, regardless of their background, have equal access to resources, opportunities, and a fair treatment. This chapter will delve into the importance of social equity, how it can be built into a changing world, and the role that user interactions play in contributing to this change. We will also briefly discuss about how AI can play a major role in this aspect. Traditionally, the topic of social equity has either been neglected or misinterpreted and misrepresented by different cohorts of people. To add value to the conversation of social equity and its significance, we shall need to understand some of the basic concepts and principles on this aspect.

Social Equity and the Need for It

In a world that is rapidly becoming more interconnected and diverse, social equity is essential to create a just and inclusive society. It involves the equitable distribution of resources, opportunities, and power, which is vital for creating and maintaining social cohesion. The importance of social equity cannot be overstated for several reasons:

© Raghu Banda 2024
R. Banda, *Building Social Equity with AI*, https://doi.org/10.1007/979-8-8688-0091-7_1

- **Promoting social justice**: Social equity is the foundation of a fair society. It ensures that everyone, regardless of their background or socioeconomic status, has an equal opportunity to succeed and thrive.

- **Enhancing economic growth**: Studies have shown that societies with a higher degree of social equity tend to experience more significant and sustained economic growth. This is because a more equitable society can harness the full potential of its human capital, leading to increased productivity and innovation.

- **Reducing social tensions**: When social equity is upheld, people are less likely to feel marginalized, which can reduce social tensions and contribute to a more harmonious society.

Beyond the aforementioned factors, there exist a multitude of compelling arguments for the heightened necessity of social equity in today's rapidly evolving global landscape.

The concept of social equity is frequently conflated with related notions of social justice, social norms, racial equity, and identity politics. This misinterpretation underscores the necessity of distinguishing the subtle nuances between these interconnected topics. Primarily, it is essential to delve into the precise definition and understanding of social equity.

Social equity represents the contribution made by individuals, corporations, or other entities toward the betterment of society or the system they inhabit, rather than what they extract from it. Simply put, it emphasizes the construction of social equity within the system to promote a seamlessly functioning society.

A harmonious and well-functioning society yields substantial benefits. As we stand on the precipice of a future increasingly influenced by artificial intelligence (AI), the significance of social equity becomes amplified. In an AI-dominated world that threatens to become robotic and monotonous, comprehending the role and implications of social equity in relation to our surrounding world becomes indispensable.

Building Social Equity into a Changing World

Achieving social equity in a changing world requires concerted efforts from all stakeholders, including governments, businesses, and individuals. Here are some strategies that can help build social equity:

- **Implement inclusive policies**: Governments should enact policies that ensure equal access to resources and opportunities for all citizens based on skills, experience, and merit. This may include affirmative action policies, income redistribution programs, and equal access to education and healthcare.

- **Encourage corporate social responsibility**: Businesses have a significant role to play in promoting social equity. They can do so by adopting fair labor practices, supporting local communities, and implementing diversity and inclusion policies in their workforce, at the same time not losing their core responsibility to the shareholders.

- **Foster community engagement**: Individuals can contribute to social equity by engaging in local community initiatives, volunteering, and advocating for marginalized groups coupled with incentives from corporations and governments.

- **Promote education and awareness**: Raising awareness about the importance of social equity and fostering a culture of inclusivity can help create a more equitable society.

In the preceding discussion, we have highlighted several strategies aimed at integrating social equity into our dynamically evolving world. However, beyond these, there exist numerous external and internal factors which contribute to this vital process. Among these, particularly noteworthy are the quantifiable interactions and transactions between individuals, corporations, and broader entities. These interactions can morph into a multitude of forms, each illustrating a unique method of embedding social equity into the transforming world.

In the context of the contemporary landscape, as penned post the debut of Generative AI on the global scene, the role of AI becomes particularly salient. The facets of artificial intelligence significantly influence the ways we can incorporate social equity into the metamorphosing world. AI's ability to perceive, understand, and possibly even predict human behavior without any bias allows it to present a novel approach to fostering an equitable society in a rapidly changing world.

In the following sections, we will further explore various aspects of social equity, providing in-depth analysis, case studies, and practical examples that demonstrate how it can be achieved and maintained in a constantly changing landscape.

The Role of Government in Promoting Social Equity

Governments play a critical role in advancing social equity through the implementation of policies and programs designed to level the playing field for all citizens.

Some examples of government initiatives to promote social equity include the following:

Progressive taxation systems: These systems aim to reduce income inequality by taxing higher-income earners at a higher rate than lower-income earners. The revenue generated can be used to fund social programs that benefit disadvantaged groups while ensuring that the system is not being taken advantage by the underprivileged groups.

Universal healthcare and education: Ensuring equal access to healthcare and education is fundamental to achieving social equity. Governments can invest in public healthcare and education systems that provide high-quality services to all citizens, regardless of their ability to pay. Tax incentives for healthcare professionals and education in higher education sector for any volunteer services could also lead to lower cost of healthcare and education and indirectly foster social good at the grassroots level and lead individuals to adopt a social good mindset.

Social safety nets: Governments can establish social safety nets that provide financial support to individuals who are unemployed, disabled, or otherwise unable to work. This can include unemployment benefits, disability pensions, and income support for families with dependent children. To elaborate, financial support to unemployed could be tied to individual's self-development to fit into the changing workforce.

Affordable housing initiatives: Governments can invest in affordable housing programs to ensure that all citizens have access to safe and stable housing. This can include public housing, rent subsidies, and incentives for private developers to construct affordable housing units.

The Role of the Private Sector in Promoting Social Equity

The private sector can also play a significant role in promoting social equity by adopting responsible business practices and investing in initiatives that contribute to a more inclusive society.

Some ways in which businesses can support social equity include

- **Implementing diversity and inclusion policies:** Companies can create diverse and inclusive workplaces by implementing policies that promote the recruitment, retention, and advancement of employees from diverse backgrounds based on skills and merit. This can include diversity training, mentoring programs, and flexible working arrangements.

- **Providing equal pay and benefits**: Equal work does not necessarily equate to equal quality of work. Therefore, companies should adopt strategies to measure and quantify the quality of work before implementing equal pay policies. Additionally, market demand for certain job functions and skills often dictates pay rates. Hiring individuals based on market conditions, even if it violates corporate policy, can create friction among employees and lead to internal corporate gossip. To maintain fairness and transparency, it's crucial for companies to clearly communicate their rationale and ensure that pay structures are aligned with both quality and market demands.

- **Supporting local communities**: Businesses can invest in local communities by providing jobs, sponsoring community events, and supporting local charities and nonprofit organizations.

- **Supply chain responsibility**: Companies can promote social equity by ensuring that their supply chains adhere to ethical labor practices and environmental standards.

Intersectionality and Social Equity

Intersectionality is a concept that recognizes that individuals can experience multiple and interconnected forms of oppression and privilege based on factors such as race, gender, sexual orientation, and socioeconomic status. Acknowledging intersectionality is crucial for achieving social equity, as it helps to identify and address the unique challenges faced by individuals who belong to multiple marginalized groups.

For instance, a woman of color may face discrimination based on both her gender and race, which can result in a unique set of barriers to accessing resources and opportunities. By considering intersectionality in the development of policies and programs, governments and organizations can more effectively address the complex and interconnected forms of inequality that individuals may experience.

Measuring Social Equity: Indicators and Metrics

We will discuss this in depth in Chapter 6 with the metrics and measurement. Evaluating the progress made in achieving social equity is essential for identifying areas of improvement and informing future policy decisions. Some key indicators and metrics that can be used to measure social equity include the following:

> **Income distribution**: The distribution of wealth within a society is a key indicator of social equity. Metrics such as the Gini coefficient and the income share of the top 10% and bottom 10% of earners can provide insight into income inequality.

> **Education attainment**: Educational attainment is closely linked to social equity, as it influences an individual's access to resources and opportunities. Metrics such as high school graduation rates, college enrollment rates, and literacy rates can be used to assess progress in this area.

Though perhaps not as frequently deliberated with equal emphasis, the principles of meritocracy and the consequential impact of merit and quality are integral components in the construction of social equity. At times, individuals possessing the requisite skills and potential for advancement, perhaps belonging to a forward caste, or of white ethnicity, or identifying as male, may face constraints due to the very aspects of their identity such as race, sex, or gender.

This situation calls for a reframing of our perception, prompting us to view these circumstances through the lens of merit and quality. By recognizing and rewarding individuals based on their abilities and performance, irrespective of their social or demographic characteristics, we can foster a more equitable society. In this manner, meritocracy can serve as an effective means of ensuring fair opportunities for all, promoting a more balanced way forward in the face of sociocultural diversity.

Tables 1-1 and 1-2 present a bit more detail by providing a tabular representation of some of these facts.

Table 1-1. *Global Gini Coefficients*

Country	Gini Coefficient
Country A	0.45
Country B	0.38
Country C	0.32
Country D	0.47
Country E	0.29

Note: The Gini coefficient measures income inequality within a country. A Gini coefficient of 0 represents perfect equality, while a Gini coefficient of 1 represents maximum inequality.

Table 1-2. *Gender Pay Gap by Industry*

Industry	Male Median Salary	Female Median Salary	Pay Gap
Industry A	$50,000	$42,000	16%
Industry B	$60,000	$54,000	10%
Industry C	$45,000	$40,000	11%
Industry D	$55,000	$49,000	11%
Industry E	$65,000	$58,000	11%

Note: The pay gap is calculated as the percentage difference between male and female median salaries in each industry.

Case Studies

Case studies can provide valuable insights into the practical application of social equity principles and serve as inspiration for future initiatives. Here are two brief examples of successful social equity initiatives.

Case Study 1: Affordable Housing Program in City X

City X implemented a comprehensive affordable housing program that involved partnering with private developers, offering incentives for the construction of affordable units, and expanding public housing options. As a result, the city experienced a significant increase in affordable housing availability, contributing to a more equitable distribution of resources among its residents.

Case Study 2: Inclusive Education Initiative in Country Y

Country Y launched an inclusive education initiative aimed at ensuring that all children, regardless of their background, have access to quality education. The initiative involved training teachers in inclusive education practices, providing resources and support for students with disabilities, and developing culturally sensitive curricula. This initiative led to increased enrollment and retention rates among marginalized students, promoting greater social equity in education.

The Changing World Order

The changing world refers to the ongoing transformations in technology, globalization, and societal norms that are reshaping the way we live, work, and interact with one another. User interactions, particularly through digital platforms, have become increasingly influential in driving these changes.

Some of the ways in which user interactions contribute to the changing world and social equity include the following:

Breaking down barriers: Digital platforms have made it easier for people from different backgrounds to connect, share ideas, and collaborate. This increased connectivity can help break down cultural, social, and economic barriers and contribute to a more equitable world.

Amplifying marginalized voices: Social media and other digital platforms have provided marginalized groups with a means to make their voices heard and raise awareness about their struggles. This has helped in fostering empathy and understanding among people from different backgrounds, thus

promoting social equity. In some scenarios,
these broken barriers in fact are leading to social
inequality wherein work is being outsourced,
leaving more people unemployed and more
disparity in financial status. We discuss this in detail
in the later chapters.

Facilitating social movements: User interactions
have played a crucial role in mobilizing support
for various social movements aimed at promoting
social equity, such as the Black Lives Matter
movement and the Me Too movement. These
platforms have enabled people to come together in
pursuit of a more just and equitable world.

Now, let us unpack a bit further as we traverse the path of the 21st
century and discuss how the world order is experiencing transformative
shifts catalyzed by rapid advancements in technology, specifically through
digitization, modernization, and the influence of artificial intelligence,
including Generative AI. These phenomena are not just changing the way
we work, interact, and perceive our surroundings but are also significantly
impacting the architecture of social equity within our global system.

Digitization: The process of converting
information into a digital format has revolutionized
communication and accessibility of information.
With the world now interconnected via a vast digital
network, individuals regardless of their geographic
location have unprecedented access to resources,
knowledge, and opportunities. This democratization
of information has the potential to level the playing
field, promoting social equity by diminishing the
barriers of distance, socioeconomic status, and even
language.

Modernization: Defined by the adoption of new technologies and the transformation of traditional systems, modernization is instrumental in mitigating social inequities. By streamlining processes, improving productivity, and expanding reach, modern technologies can reduce the constraints traditionally associated with factors such as age, gender, disability, and socioeconomic status. This broader inclusion creates a more equitable world, encouraging participation from diverse groups and ensuring that their voices are heard.

Artificial intelligence (AI): AI is arguably the most disruptive and influential technological development of our time, playing a decisive role in shaping social equity. AI and machine learning algorithms can analyze patterns and trends within large datasets, highlighting disparities and biases that may otherwise go unnoticed. Furthermore, AI can offer solutions to address these inequities, potentially reshaping social, economic, and political structures toward a more equitable model.

The advent of Generative AI: Capable of creating content that is convincingly human-like, GenAI introduces a new dimension in this discourse. By simulating human-like content and decision-making processes, Generative AI can help in understanding and addressing human biases. It could also ensure fair representation in the digital world, promoting diverse narratives and voices. Moreover, in the realm of education and job training, Generative AI can provide personalized learning experiences and skill development, further fostering social equity.

However, it is crucial to acknowledge the double-edged nature of these technologies. While they offer promising pathways for promoting social equity, they could inadvertently deepen disparities if not managed with deliberate care. As such, the ethical implementation of these tools becomes paramount to ensure the benefits are reaped by all, and not just a privileged few.

Let us go back and dive a bit into the user interactions in the context of this changing world order and how we are undergoing a significant transformation. As digital platforms proliferate, the ways in which individuals communicate, work, learn, and participate in society have become increasingly mediated by technology. The pervasiveness of these platforms implies that our digital actions, decisions, and interactions leave behind a rich trail of data. These digital footprints, when aggregated, can offer remarkable insights about us as individuals, communities, and as a society at large.

Artificial intelligence and Generative AI have a crucial role to play in processing and understanding these vast datasets. AI, with its capacity for pattern recognition and predictive modeling, can identify trends, predict outcomes, and even proactively influence interactions based on these insights without any human emotions. This has significant implications for social equity.

On a positive note, AI can uncover and quantify systemic biases within these digital interactions that might otherwise go unnoticed. For example, AI algorithms can assess the language used in job advertisements for potential gender or racial biases or evaluate lending practices for signs of economic discrimination. By illuminating these biases, AI enables us to address them directly, creating more equitable systems.

Generative AI, on the other hand, goes a step further. By simulating human-like content and decision-making processes, it can generate new and diverse narratives. For instance, in digital platforms, Generative AI can generate personalized content, ensuring that a variety of perspectives are presented and promoted. This technology can help ensure fair representation in the digital world, amplifying marginalized voices and providing a platform for diverse narratives.

AI and Generative AI can also contribute to education equity. Personalized learning platforms, powered by these technologies, can provide customized learning experiences tailored to each individual's unique needs and pace. This democratization of education could bridge the knowledge divide, offering an equitable chance at quality education for all, regardless of geographical or socioeconomic boundaries.

Nevertheless, as these technologies shape user interactions and influence social equity, we must stay cognizant of their inherent risks. Unchecked, AI can perpetuate and amplify existing biases in the data it's trained on, leading to discriminatory outcomes. It is imperative that the development and deployment of these technologies be guided by robust ethical frameworks that prioritize fairness, transparency, and inclusivity.

Overall, the evolving world order characterized by digital user interactions, under the influence of AI and Generative AI, provides both significant opportunities and challenges for social equity. By leveraging these technologies responsibly and ethically, we can harness their potential to build a more equitable world.

In sum, the changing world order spurred by digitization, modernization, and AI, is making a profound impact on building social equity into our global system. The road ahead requires careful navigation to ensure these tools are wielded not as engines of division but as catalysts for a more equitable and inclusive world.

Conclusion

In this chapter, we focused on the aspects of social equity at a high level and understood the key concepts around how social equity is built into the system and the various factors that go into it. We also briefly analyzed how AI and GenAI can play a big role in this. In the next chapter, we shall dive into the core aspects of equality and equity through the lens of AI.

In the course of this chapter, we have embarked on a comprehensive exploration of social equity, scrutinizing the fundamental concepts that underpin its integration into our societal systems. Our discussion extended beyond the primary factors contributing to social equity, to deliberate on the potential role of artificial intelligence, including Generative AI, as transformative agents in this context.

While we have embarked on an initial examination of how AI technologies could profoundly influence the construction of social equity, this chapter represents only a high-level overview of these complex interrelations. We remain cognizant of the breadth and depth of this multifaceted issue, as well as the vast potential that lies in the further application of AI.

As we transition into the forthcoming chapter, we will shift our focus to delve deeper into the nuanced intricacies of equality and equity, particularly through the perspective of AI. Here, we aim to unravel the finer details of how AI can be harnessed to foster a more just and equitable world, offering fresh insights into the pivotal role of these technologies in shaping our future society.

CHAPTER 2

Equality and Equity

In the rich tapestry of societal evolution, the threads of equality and equity stand out, not merely as terminologies but as guiding principles of human endeavor. On the surface, these words, often whispered in the same breath, seem synonymous. Yet, diving deeper, they unfurl distinct philosophical narratives and responsibilities.

As we navigate this chapter, we shall traverse the nuanced contours differentiating equality from equity, seeking not just definition but deep-rooted understanding. We'll juxtapose these concepts against the backdrop of people, the frameworks of processes, and the ever-evolving canvas of technology. But it's not merely an intellectual exploration; it's a journey of the soul, a quest to discern the ethos that shapes our collective future.

And as we chart this course, let's take a solemn moment to acknowledge the unsung custodians of social equity. These quiet warriors, often overshadowed by louder narratives, are the torchbearers ensuring that the balance between equality and equity doesn't waver in the tempests of time!

In our digital age, seemingly fueled by relentless ambition and innovation, we are summoned to an introspective juncture: the intricate balance between equality and equity. As we stand on the precipice of a new era, these guiding principles don't just anchor our technological endeavors; they serve as beacons, illuminating the vast moral expanse before us.

© Raghu Banda 2024
R. Banda, *Building Social Equity with AI*, https://doi.org/10.1007/979-8-8688-0091-7_2

Historically, the tapestry of human interaction was woven with threads of tangible exchanges. But today, AI, and more pointedly, Generative AI, has ushered in a new paradigm. This isn't merely technology in action – it's a profound reflection of our collective aspirations, biases, and shared humanity. In this transformative age, the constructs of equality and equity are not static markers, but dynamic compasses, guiding us through an ever-evolving ethical maze.

Beyond the captivating allure of technology lies a deeper, timeless narrative, echoing with the wisdom of generations past. Interwoven are themes of ethics, sustainability, privacy, and our shared responsibilities as custodians of this digital epoch. As we grapple with these twin pillars of equality and equity in our technocentric world, we are penning a philosophical legacy – one that will be pored over, scrutinized, and perhaps revered by future thinkers and innovators.

What Is Equality and What Is Equity?

In a world that is rapidly becoming more interconnected and diverse, social equity is essential to create a just and inclusive society. It involves the equitable distribution of resources, opportunities, and power, which is vital for creating and maintaining social cohesion. The importance of social equity cannot be overstated for several reasons.

In the grand tapestry of human society, two ideals often stand juxtaposed: equality and equity. While these terms are sometimes used interchangeably, their nuanced differences are crucial to understanding the dynamics of human progression.

Equality, at its core, espouses a universalist approach. It hands out the same tools to everyone, upholding the principle of undiscriminating access. But in doing so, it presumes a level playing field, which, when we delve deep into the interwoven complexities of human existence, might be more of an aspiration than a reality.

Equity, however, delves deeper. It seeks not just to provide, but to understand. It contemplates the uneven terrains of personal history, cultural heritage, and inherent challenges that each individual faces. It's not about treating everyone the same, but about ensuring that everyone can arrive at the same endpoint. Equity tailors its approach, acknowledging that fairness may sometimes require differentiated treatment.

Picture, if you will, a baseball game or a game of cricket – an embodiment of competition, skill, and passion. Now, imagine three spectators, each a unique testament to nature's diversity in height. Handing them an identical platform (equality) might seem just, but it fails in the practical realm where the shortest might still be deprived of the joy of the game.

Equity would instead mold the solution, granting platforms of varying heights to ensure every spectator enjoys the game in its full vibrancy.

To truly advance as a society, it's pivotal we understand these nuances. It's about progressing from blanket solutions to tailored interventions – from equal provisions to equitable outcomes.

In our intricate web of human interactions, the concepts of equality and equity often emerge as guiding beacons. However, their connotations are markedly distinct, and discerning their subtleties is key to effective social progress.

At first glance, equality is straightforward. Consider a classroom where each student receives the same textbook. This is equality in action: uniformity in resource distribution. But, what if one student is visually impaired? The same book, though equally distributed, doesn't offer equal value.

Enter the realm of equity. Here, the visually impaired student might receive a Braille version or an audiobook. It's a tailored response, adjusting resources to meet individual needs. But, it's essential to understand that equity isn't exclusively outcome-focused. The essence of equity lies as much in the processes, considerations, and metrics of measurement as it does in the final outcomes.

For instance, in a marathon, the outcome-focused approach would be to ensure every runner finishes at the same time – a flawed and impractical metric. A more equitable process-oriented approach ensures that every participant has equal access to training, nutrition, and gear. The outcome, the race's finish, varies, but the equity lies in the access and opportunity.

Another compelling example lies in our educational systems. Equity isn't merely about ensuring all students score identically on a test. It's about ensuring each student has access to the resources, tutoring, and environment they need to succeed, given their unique challenges. It's also about recognizing that success isn't one-dimensional. A standard test might measure memory or analytical ability, but what about creativity, emotional intelligence, or practical skills? True equity appreciates the multiplicity of talents and seeks diverse metrics to honor them.

In essence, while equality offers a one-size-fits-all solution, equity dives into the complexities of individual experiences. It acknowledges that fairness isn't just about uniformity in results; it's also about the integrity of the journey and the metrics by which we gauge success.

Addressing Equality and Equity in the Context of People, Processes, and Technology

As we are discussing, in the intricate dance of modern society, we're beckoned by two guiding lights: equality and equity. These aren't mere ideals, but essential touchstones that influence every facet of our existence, from the individual narratives we weave and the systems and processes we architect to the groundbreaking technologies we champion. As we stand on the precipice of an era marked by unprecedented change, it's imperative to understand and harmonize these principles, ensuring that our strides toward progress are both universally inclusive and uniquely tailored.

With this foundation set, we can then delve deeper into each of the aforementioned domains – people, processes, and technology!

People

In this vast tapestry of our shared human narrative, the quest for equality and equity is pivotal. However, it's essential to recognize that merit and quality, too, are integral to this discourse. Our journey is not merely about leveling the playing field but ensuring that the game played is of the highest standard.

Take Sarah, a first-generation immigrant student, as an example. While she grapples with cultural adjustments and language barriers, she also brings with her unique insights, perspectives, and a tenacity molded by her experiences. Offering her an environment that values diversity is crucial, but equally important is acknowledging her merit. Sarah's success should be seen not just as a triumph of equity but also as a testament to her inherent quality and capability.

Transparent communication stands at the core of this journey. By creating spaces where individuals can voice their unique needs and aspirations, we can ensure a more holistic approach – one that recognizes merit and quality while also addressing issues of equity.

It's imperative to remember that embracing diversity is both a moral and a strategic endeavor. While varied perspectives undoubtedly lead to richer discussions and innovations, the quality of those perspectives, grounded in expertise and merit, is what truly propels society forward.

Resources and their strategic deployment are essential. Still, they should be viewed as catalysts that enable individuals of merit to shine brighter. Instead of merely leveling the field, our aim should be to elevate it, ensuring that excellence is recognized, nurtured, and celebrated.

To sum up, in our pursuit of a world that upholds equality and equity, it's crucial to weave in the threads of merit and quality. This ensures a society where every individual, regardless of background, is not just given an opportunity but is also inspired to rise to their highest potential, contributing to a brighter collective future!

Processes

Now at this crossroads of societal progress and business innovation, the incorporation of equality and equity in our systems is not merely a nod to ethics; it's a strategy as nuanced as configuring an SAP or a Salesforce module. These systems, while ensuring a wide-reaching impact, also place a premium on efficiency and effectiveness. Similarly, while our policies and processes must account for varied needs, they should also ensure that quality and merit are not compromised.

Consider how enterprise systems like SAP prioritize scalability without losing sight of individual user requirements. In the same vein, ensuring equal access to resources and opportunities should not just be about quantity but also about quality. Offering financial assistance to low-income students or professional development opportunities for employees from diverse backgrounds must be calibrated to their merit and potential, ensuring that resources aren't just spread wide but deep, yielding genuine returns.

When we talk about developing inclusive policies, think of it as customizing an Oracle suite for a business. Generic solutions can only go so far! Tailored solutions, which take into account the unique architecture of each department (or in this case, demographic), create systems that truly resonate. Policies should be as dynamic and adaptive as the real-time adjustments made in Salesforce dashboards, allowing for real-time modifications based on changing societal variables. This dynamism, while promoting equality and equity, should also weigh the qualitative impact and the merit of the beneficiaries.

Monitoring and evaluating progress can be likened to the performance analytics of these enterprise systems. Just as businesses don't settle for outdated modules but upgrade to enhance efficiency, societal policies must be revisited, assessed, and recalibrated. This ensures that the drive for equity doesn't eclipse the pursuit of excellence.

Building a society that champions equality and equity, while respecting merit and quality, is akin to deploying a sophisticated enterprise solution like SAP, Salesforce, or Oracle. It's about crafting a responsive system that is as inclusive as it is efficient and as equitable as it is excellent. While this comparison may seem exaggerated – comparing enterprise systems to processes tailored for people – the point is that data plays a crucial role in shaping these processes. Just as enterprise systems rely on data to optimize operations, societal systems must leverage data to ensure fairness, inclusivity, and excellence.

Technology

Coming to the vast confluence of societal dynamics and technological progress, technology stands as both a beacon of hope and a potential challenge. Like the dualities we often encounter in our narratives, technology, too, wields the power to either bridge or widen disparities.

Designing with inclusivity at the core: In the realm of technology, especially AI, there's a subtlety that's often overlooked: the DNA of its design. When developers dive deep into their codes and algorithms, they're not merely solving for an output; they're determining whose voice gets amplified and whose might get inadvertently silenced. There are numerous examples from our social media platforms like Twitter, Facebook (Meta), and more in this regard!

Take, for instance, AI-based recommendation systems in applications. Without a conscious emphasis on avoiding biases, they can unintentionally perpetuate stereotypes. However, if crafted with care, the same systems can introduce users to a world of diverse content, breaking silos and broadening horizons. Reflect on AI systems like those employed by major corporations such as Google or Microsoft. These platforms could, if designed with a narrow perspective, skew business decisions toward certain demographics, missing out on diverse opportunities. Conversely, when developed with an inclusive lens, they could lead industries toward untapped potentials and markets, emphasizing both quality and representation.

Bridging the digital chasm: Digital divides aren't merely about technology; they're reflections of deeper societal rifts. Consider rural areas with limited Internet access. For them, the digital revolution is more a distant dream than an immediate reality. But just as AI tools like deep learning can extract patterns from vast datasets, governments and organizations can harness insights to strategize targeted interventions.

For example, AI-driven analytics can identify regions with stark digital literacy disparities. These insights can inform policies, much like a business strategy is refined based on market analysis, ensuring resources such as Internet connectivity and digital education are channeled effectively.

Tech as an altruistic force: In the annals of tech history, the chapters that truly resonate are those where technology stands as a beacon for the betterment of society. AI is not just about optimization; it's about transformation.

Consider AI-enhanced platforms that mirror the efficiency of systems like SAP but are tailored to address societal challenges. Whether it's an app designed to monitor and promote mental health or a platform that bridges volunteers with community endeavors, the onus is on ensuring that while we chase innovation, we don't compromise on the essence of humanity.

Moreover, AI tools, when designed with precision, can unearth deep-rooted societal disparities. The power of deep learning could, for instance, analyze vast arrays of data to pinpoint where inequalities are most pronounced, guiding policy and action.

The triad of excellence – merit, quality, and equity: In the grand scheme of technological advancement, the compass must point toward three cardinal directions simultaneously: merit, quality, and equity. The goal is not just to democratize technology but to elevate its standard. As we engineer platforms that cater to diverse audiences, the gold standard remains – ensuring top-tier quality.

In the confluence of technology's promise and society's needs, the narrative should champion a dance where innovation meets inclusivity without missing a beat on merit.

Recognizing Unsung Heroes of Social Equity

Navigating the ever-evolving landscape of our global society, it's evident that alongside the champions of social equity, merit and quality play pivotal roles. As businesses grow and societies progress, a balanced emphasis on all these facets can create a robust blueprint for a more harmonious future.

There's a vibrant momentum of individuals and groups championing social equity. While many champions remain unsung, their stories, when interwoven with the importance of merit and quality, create a compelling narrative of progress.

Highlighting real-life champions: The Pakistani education activist Malala Yousafzai didn't just advocate for education; her resilience, eloquence, and conviction showcased merit. This with her unwavering voice made her stand out. Platforms ranging from documentaries to articles can echo similar tales, highlighting that quality and merit are as crucial as the struggles against societal barriers.

Recognizing excellence and dedication: Awards like the Goldman Environmental Prize don't just spotlight activism; they spotlight exceptional activism. It also emphasizes the importance of impactful, quality work on ground. This dual emphasis ensures we're not just uplifting voices but uplifting voices that drive change through quality efforts. By emphasizing both merit and the cause, we broaden our scope of recognizing and uplifting real change-makers.

Building bridges through collaboration: Platforms such as TEDx have become revered not merely for the variety of topics they showcase but for the caliber of their speakers, emphasizing both substance and expertise. These kind of platforms curate quality content, blending the importance of merit with the cause, ensuring the narratives that emerge are both inspiring and of substance.

Now let's peel the layers with examples around specific sectors like education, healthcare, and the workplace a bit further:

Education: Khan Academy isn't just revolutionary due to its reach but also its quality. The emphasis isn't merely on making education accessible but ensuring it's of top-tier quality, tailored to individual needs.

Healthcare: Platforms like Zocdoc and NexHealth have democratized healthcare access. But it's not just about bridging distances. The emphasis is also on quality care, ensuring patients, irrespective of their location, get top-notch consultations. These platforms' growth isn't just due to their innovative approach to healthcare access but their emphasis on connecting patients with qualified professionals ensuring that healthcare quality remains uncompromised.

Workplace dynamics: When Salesforce tackled the gender pay gap, it wasn't just a pursuit of equity. It was also about recognizing and rewarding merit and quality work, ensuring every employee, regardless of their gender, received deserving compensation.

Let's now discuss how tracking progress requires a multifaceted approach with some examples in different sectors:

Socioeconomic metrics: The Grameen Bank's microfinancing isn't just about loans; it's about backing individuals with potential and merit, changing socioeconomic dynamics based on quality and capability.

It isn't just pioneering because of its unique lending model but because it's a beacon for how backing meritorious individuals can result in quality ventures that uplift entire communities.

Education measures: Teach For America isn't just about filling teaching slots but ensuring quality education by placing meritorious graduates in underserved regions.

Teach For America's strategy is brilliance twofold – filling educational gaps and ensuring that these gaps are filled by talented, dedicated professionals.

Health metrics: Targeted interventions, like polio eradication campaigns, succeeded not just due to widespread coverage, but also because of the quality of the healthcare provided. Campaigns targeting healthcare issues succeed not just through outreach but due to the quality of healthcare rendered.

Workplace metrics: When Intel aimed for a diverse workforce, it wasn't just about numbers. It was about bringing quality, varied perspectives to the table, ensuring innovation stemmed from merit. A drive for a diverse workforce looked beyond mere representation. The focus was on harnessing quality, diverse perspectives to foster innovation.

As we navigate the fusion of AI and data, discernment remains key. AI, while a tool of precision, should also be a beacon of quality and merit, ensuring its outputs are as exemplary as its algorithms. Incorporating AI and data analytics into this narrative, we're reminded that these tools, as revolutionary as they are, must uphold standards of quality and reflect the merits of unbiased data processing.

In this vast realm of understanding equality and equity, sometimes a simple visual or quantitative insight can convey profound truths. Such data representations not only elucidate disparities but also inspire targeted solutions.

Equity and Equality Mixed with Merit and Quality

In this section, we'll delve into some examples that mix merit and quality.

Diverging Educational Outcomes Across Socioeconomic Tiers

Table 2-1 underscores the significant disparities across socioeconomic backgrounds. High school graduation and college enrollment rates starkly contrast between low-income and high-income students, highlighting a tangible area for educational equity reforms.

Table 2-1. *Education Indicators by Socioeconomic Background*

Indicator	Low-Income Students	Middle-Income Students	High-Income Students
High school graduation rate	75%	85%	95%
College enrollment rate	35%	60%	85%
College graduation rate	20%	50%	75%

Note: This table shows how educational outcomes can vary significantly based on a student's socioeconomic background, highlighting the need for greater equity in education.

Navigating the corridors of education isn't just about securing that degree; it's about unlocking the door to a brighter future, in the form of meaningful employment. Let's weave this into our examination, underscoring the tangible outcomes of our education indicators, by dissecting the complex layers of equality, equity, merit, quality, and real-world opportunities.

High School Graduation Rate

Equality: Every student should have the unbridled chance to graduate, and with that, the access to entry-level jobs or internships. Think of the teenager from a small town landing a hands-on apprenticeship with a local mechanic after graduation.

Equity: Some students might require vocational courses, parallel to mainstream education, catering to their unique talents, ensuring they too get employment opportunities. A case in point: offering career-focused workshops in arts or crafts for those inclined.

Merit: Top performers should have the gateway to premium internships, like a student with a knack for tech joining a startup's coding team right out of school.

Quality: But, ensuring that every student is employable, irrespective of the path they choose, is the real testament to an education's quality.

College Enrollment Rate

Equality: Enrolling in college should offer a wider canvas of job opportunities post-graduation.

Equity: Financial aid or campus mentorship programs can ensure that socioeconomic background doesn't dictate one's job prospects. Consider the college that partners with industries, ensuring internships for students from all economic backgrounds.

Merit: Elite programs and honors courses can open doors to prestigious job placements. A top-notch finance student might land an internship at a Wall Street firm.

Quality: Beyond enrollment numbers, the depth and breadth of campus placement opportunities are markers of an institution's commitment to quality.

College Graduation Rate

Equality: Graduating college should ideally mean a higher shot at enhanced job roles and better pay.

Equity: Tailored campus placement programs can help bridge gaps, allowing marginalized groups the same shot at plum jobs, like a mentorship program specifically for first-generation college students.

Merit: Excellence should open elite avenues – think of being headhunted by the likes of Google or SpaceX due to outstanding academic performance.

Quality: Finally, if a degree doesn't translate to job readiness or lacks industry relevance, then it questions the essence of quality in education.

The real potency of an education system isn't just in the classroom but in its ripple effect – in how it shapes careers and life trajectories. Balancing equality, equity, merit, and quality isn't a simple task, but when done right, it paves the way for fulfilling, prosperous lives for its students.

Healthcare Disparities Among Demographics

As we peer into Table 2-2, the gaps in healthcare accessibility among different demographic groups become strikingly clear. It's a testament to the need for more nuanced, equity-focused healthcare policies.

Table 2-2. *Healthcare Access by Demographic Group*

Demographic Group	Percentage with Health Insurance	Percentage with Regular Access to Primary Care
Group A	90%	80%
Group B	75%	65%
Group C	85%	70%
Group D	95%	85%

Note: This table illustrates disparities in healthcare access between different demographic groups, emphasizing the need for targeted interventions to promote greater equity in healthcare.

When discussing healthcare access by demographic groups, it's essential to understand the nuances that lie beyond simple statistics. Let's navigate through the intricate mosaic of equality, equity, merit, and quality in this realm.

Healthcare Access by Demographic Group

Equality: At the very core, every individual, irrespective of their demographic identity, should have access to basic healthcare. Imagine a scenario where every city resident, from the downtown business executive to the artist in the uptown loft, receives the same baseline health services.

Equity: Different demographic groups might have unique health needs. For instance, a community living near an industrial zone may be at higher risk for respiratory ailments and thus might need more specialized care. Similarly, the elderly might require more frequent health screenings. Providing tailored services to meet these specific needs ensures that everyone gets an equal shot at optimal health.

Merit: Merit in healthcare might sound a bit unconventional but think of it this way – those professionals and researchers who excel in their medical pursuits should be given the tools and platforms to innovate and contribute more extensively. For instance, a doctor who has made significant strides in pediatric care in an underserved community might merit additional funding or resources to magnify their impact. This ensures that healthcare improves holistically, benefiting various demographic groups!

Quality: Access alone isn't enough; it must be timely quality access. It's not just about how many people get healthcare, but how effective and efficient that healthcare is. Are treatments leading to better health outcomes? Are diagnoses accurate? For instance, if two different demographic groups have equal access to cancer screening but one group consistently gets earlier and more accurate diagnoses due to better equipment or training, there's a clear quality gap.

At a glance, the equation seems simple: everyone should have equal access to healthcare. But dig a little deeper, and the intricacies of equity, merit, and quality emerge, challenging the narrative of a one-size-fits-all solution.

Urban vs. Rural: An Insight into Healthcare Access

Here's a practical example to tie it all together: let's consider two demographic groups, Group A living in urban areas and Group B in rural areas!

Urban Landscapes (Group A)

Imagine the hustle and bustle of New York City – skyscrapers, diverse demographics, and an ever-growing population.

Equality: The ideal scenario is akin to a well-oiled machine where every New Yorker, from the Wall Street executive to the artist in Brooklyn, has a clinic or hospital within a reasonable distance.

Equity: But some boroughs, particularly with industrial zones, might grapple with heightened pollution levels. Specialized respiratory clinics in such areas don't just make sense; they're a need.

Merit: Enter the movers and shakers of the medical world. Those researchers making groundbreaking discoveries in urban health should be armed with the resources to drive their vision. Imagine if a young researcher from Columbia University found a novel way to combat urban air pollution–related ailments, such merit demands backing.

Quality: New York prides itself in being the best. So why should its healthcare be any different? State-of-the-art clinics, continuous medical training, and patient-centered care should be the gold standard, not the exception.

Rural Settings (Group B)

Now, transport yourself to the serene landscapes of rural Montana – vast open spaces, limited population, and unique healthcare challenges.

Equality: Here, the challenge isn't just healthcare access, but accessibility itself. A Montana rancher shouldn't have to drive for hours for basic medical services.

Equity: With vast terrains come vast challenges. Mobile clinics, perhaps inspired by the resourcefulness of grassroots movements in India or Africa, could make periodic rounds, ensuring even the most isolated resident isn't left behind.

Merit: Rural healthcare, with its distinct challenges, beckons for innovative solutions. Consider a doctor in Montana pioneering telehealth solutions well before it became a buzzword, connecting her patients to specialists across the country. That's merit that can't be ignored.

Quality: And while the settings are rural, the medical standards shouldn't be. Telemedicine consultations with Mayo Clinic experts or Johns Hopkins specialists could bridge the quality gap, ensuring rural doesn't equate to secondary!

The intertwining of equality, equity, merit, and quality in healthcare, particularly when mapped to demographic contexts, has transformative potentials. By ensuring each demographic group has tailored access to quality healthcare, society as a whole stands to gain in manifold ways. Let's extrapolate the outcomes when the principles of equality, equity, merit, and quality seamlessly integrate into our healthcare fabric.

Economic Benefits

As the adage goes, "Health is wealth." By bolstering the health of the population, urban centers, like New York, can foster a more robust, active, and efficient workforce. Reduced absenteeism due to health issues means higher productivity levels, driving economic growth.

Rural regions, like Montana, experience reduced healthcare-induced financial hardships. Fewer families would face the stark choice between medical bills and daily expenses, leading to increased spending in local economies and overall economic upliftment.

Enhanced Innovation

By recognizing and supporting merit, urban healthcare challenges would become hotbeds of innovation, attracting global talent and investments. Imagine New York City becoming the global epicenter for combating urban health challenges.

Rural landscapes might inspire grassroots innovations, perhaps leading to mobile telehealth solutions that become global models. Picture Montana's healthcare solutions being replicated in rural Australia or Siberia.

Strengthened Community Bonds

When quality healthcare is accessible, urban dwellers witness reduced health disparities, fostering stronger community bonds. A Brooklyn artist and a Wall Street executive, both champions of their health, find common ground, fostering a more cohesive community.

In rural settings, the mobile clinics and telehealth services not only bring medical services but also a sense of connectivity. A farmer in Montana might, through telemedicine, be discussing his symptoms with a specialist in Baltimore, bridging gaps and fostering understanding between urban and rural worlds.

Global Leadership

With cities like New York and serene landscapes like Montana serving as models, the United States can assert its position as a global leader in healthcare innovation and delivery. By navigating its unique challenges, America can share its holistic healthcare model – one that balances equality, equity, merit, and quality – with nations worldwide.

In essence, a nation that prioritizes all these facets in healthcare is not just investing in the well-being of its people but in the very foundations of its future. Healthier individuals build more resilient communities, and these communities, in turn, forge a nation that's robust, innovative, and harmoniously interconnected. The ripple effect of such a comprehensive approach to healthcare could be the beacon for societal evolution in the 21st century.

Blending access and quality in healthcare isn't just about infrastructure or funds; it's about ethos, ambition, and relentless pursuit of excellence. From the crowded streets of urban cities to the quiet corners of rural landscapes, ensuring everyone gets not just a piece of the pie but the best slice is what equality, equity, merit, and quality, truly embodied, can achieve.

The Employment Chasm: Disability in Focus

The stark employment gap, especially for persons with disabilities, reveals a pressing concern. Table 2-3 reiterates the importance of fostering an inclusive workforce and bolstering support mechanisms for individuals with disabilities.

Table 2-3. *Employment Rates by Disability Status*

Disability Status	Employment Rate
Persons with disabilities	33%
Persons without disabilities	74%

Note: This table demonstrates the significant employment gap between persons with and without disabilities, highlighting the need for greater equity in the workforce and support for individuals with disabilities.

In a world saturated by headlines, data points, and a growing appetite for diversity, there's a pressing need to address the topic of employment rates by disability status with depth and nuance.

Equality and Equity

When we address employment rates concerning disability status, equality and equity serve distinct yet interconnected roles. Equality would argue that all individuals, irrespective of disability status, should be provided the same opportunities in the employment sector. Equity, on the other hand, understands that persons with disabilities might need specific accommodations or environments to achieve the same level of success in their careers.

Now let's consider the journey of Aisha, a software developer with a visual impairment. At face value, equality might dictate that she undergoes the same hiring process as every other candidate. But delve deeper. Equity is when Aisha is allowed to use her screen reader during coding tests, not for an advantage but to level the playing field.

Merit and Quality

Every individual brings unique skills and perspectives to a role. The metric of merit should focus on an individual's qualifications, experiences, and abilities rather than their disability status. Quality, meanwhile, pertains to the standard of work or the value addition an employee brings to a role. Persons with disabilities, when given the right tools and environment, can provide quality work at par with, if not exceeding, the work of those without disabilities.

Rahul, a wheelchair-bound architect, exemplifies merit in action. His designs are some of the most innovative in urban planning. But would he have had the chance to shine if judged solely by his physical capabilities and not the ingenuity he brought to his projects? The measure of quality is in his work's impact, not in his disability.

But Is It Fair?

Evaluating employment rates by disability status is a double-edged sword. On one hand, such evaluations shed light on discrepancies, helping societies and companies work toward inclusivity. On the other hand, it risks pigeonholing individuals based on their disability status rather than their capabilities.

Fairness emerges when such statistics are used as a starting point for discourse and improvement rather than an end in themselves. The focus should always pivot back to the individual's merit, ensuring that persons with disabilities aren't token hires but valued members of a team.

If a Silicon Valley tech giant released figures showing lower employment rates for persons with disabilities, the immediate reaction might be an outrage. However, if that data is used as a catalyst to revamp hiring processes, introduce better facilities, or even spotlight talent from within this demographic, then it serves a constructive purpose.

The Nonbinary Paradigm

The nonbinary movement, which challenges traditional notions of gender, is making waves in various sectors, including employment. When addressing employment rates by disability status, adding a nonbinary lens might seem complex but is essential for a holistic view.

Individuals who identify as nonbinary and have a disability experience a layer of potential bias and challenges. They're navigating the intersections of gender identity and disability, both of which can influence their employment opportunities. Recognizing this intersectionality is crucial for comprehensive policies.

In the current world, the confluence of disability and nonbinary identities in employment indicates the evolving nature of workplaces. Traditional metrics are giving way to more nuanced, inclusive frameworks. Companies aren't just looking for employees; they're seeking diverse voices that echo the varied tapestry of our society.

Jordan, who identifies as nonbinary and is hard of hearing, operates at the intersection of gender and disability. At face value, they may tick multiple "diversity" boxes for companies. But their true value isn't their identity tags but the blend of empathy, innovation, and resilience they bring to team dynamics.

For a world that's rapidly modernizing, there's still a conservative undertone when it comes to employment practices. Companies might boast AI-driven strategies, but human biases linger. Reducing a person to data points, be it disability or gender identity, misses the broader tapestry of their life experiences, capabilities, and the unique perspectives they can offer.

Outcomes for a Better Society

A fair society isn't one that simply acknowledges diversity but actively invests in it. By doing so, we not only amplify voices that have historically been subdued but also drive innovation and empathy at scales previously unimagined. Think of a world where Aisha's software improves user experience for millions, where Rahul's designs make urban spaces universally accessible, and where Jordan leads inclusivity workshops that transform company cultures.

Such a society isn't utopian. It's achievable. The path there is lined with respect, understanding, and a commitment to see beyond the obvious, delving into the richness of individual narratives. It's not just about making room at the table; it's about valuing the stories, insights, and brilliance each seat brings.

In conclusion, while the pursuit of equity and equality is paramount, it's incomplete without an underlying thread of merit and quality. For it's in this harmonious blend that society progresses, not just inclusively, but excellently.

In our journey toward societal progress, the pillars of equity and equality provide a solid foundation. However, the edifice truly stands strong and resilient when merit and quality are interwoven into its design. As we stride forward, by emphasizing all these aspects in unison, we carve out a pathway for a future that's not just inclusive but also distinguished and efficient.

Navigating the nuances of equality and equity, it's clear that these aren't just buzzwords but foundational pillars for societal harmony. Deep, critical insight is essential to discern their intricate differences and to articulate them in a manner accessible to all. As these concepts permeate every facet of our lives – from societal constructs to technological innovations – there's an overarching need to ensure their seamless integration with merit and quality.

There's more than just understanding these concepts. They need to be woven seamlessly into every strand of our lives, from societal frameworks and intricate processes to the pulsating heart of technology. Especially when we navigate the vast ocean of artificial intelligence, it's essential to keep the lighthouse of merit and quality in view. AI, while equipped with the potential to redefine equality and equity, should be built on the tenets of excellence and precision.

While it's pivotal to recognize disparities and strive to mitigate them, it's equally valuable to shine a light on those unsung heroes who tirelessly champion this cause. Their often-underappreciated endeavors form the very backbone of the tangible shifts we hope to actualize.

Conclusion

In culmination, our pursuit isn't just about balancing the scales of equality and equity. It's about crafting these scales with resilience and coating them with a sheen of quality. As we harness the potential of AI, let it reflect this intricate balance, embodying a narrative that's as exemplary in its function as it is in its purpose.

In wrapping up this chapter, it's evident that the discourse around equality, equity, merit, and quality is not just academic but is deeply rooted in the shared human experience. As we pivot to the upcoming chapter, our emphasis will turn toward the intersection of human science and artificial intelligence. What does it mean to truly comprehend human essence in the era of AI? And how can we leverage this unparalleled tool to not only understand but to enhance the confluence of equity and quality? The next chapter promises to be a journey into the heart of these questions, aspiring to draw a blueprint for a world that harmoniously marries technology with the intricate tapestry of human ideals.

CHAPTER 3

Human Science

In an era where digital revolution seems to redefine our existence daily, the quest to truly understand ourselves takes on a heightened significance. Human science, or as some may term, the social sciences, stands as a beacon in this endeavor. It's not merely an academic pursuit but a compass that points us toward the intricacies of human behavior, cognition, and our collective social fabric. As we chart these waters, we realize that there's a profound intersection between human science and the promises (and perils) of artificial intelligence. How does one leverage the power of AI to enrich, not undermine, our pursuit of a socially equitable world? Furthermore, in today's dynamic landscape, myriad online tools and applications offer a mirror to our social behaviors, often in ways we scarcely realize. This chapter promises a deep dive into these intersections, challenging us to see the future through a lens grounded in human understanding, all while harnessing the limitless potential of AI.

Human science, at its essence, seeks to unravel the intricacies of our behaviors, relationships, and societies. It encompasses a spectrum of disciplines, from anthropology's exploration of cultural patterns to psychology's delve into the human mind and sociology's insights into societal structures. It's a rich tapestry of understanding the "human" in every scenario, from our primeval past to our envisioned digital future.

As we stand at the precipice of the digital age, we're surrounded by tools that not only document but also influence our behavior and actions. Every click, like, share, or tweet becomes a part of the vast digital chronicle, a testament to our ever-evolving societal norms and individual identities.

© Raghu Banda 2024
R. Banda, *Building Social Equity with AI*, https://doi.org/10.1007/979-8-8688-0091-7_3

We can harness these tools to deepen our understanding of human science, to discern patterns, to predict tendencies, and to create proactive solutions.

Enter artificial intelligence – a force that, while not human, is increasingly shaping our human experiences. Its algorithms, often silent and unseen, have begun to play puppeteer to our choices. But what if, instead of passive subjects, we turn AI into our collaborator in the realm of human science? The symbiotic relationship between AI and human science can be the key to broadening our knowledge and understanding of human behavior. Instead of merely reacting to AI-driven insights, we can proactively use AI to bolster our understanding, ensuring that every algorithm, every piece of code, is imbued with a sense of human-centric ethos.

Imagine a world where AI-driven platforms don't just suggest what shoes you might like to buy next but also enhance educational equity by personalizing learning for every student. Consider digital tools that don't merely track mental health metrics but use them, in real-time, to offer meaningful interventions, breaking down barriers to mental healthcare access.

In the ensuing sections, we'll embark on this exploratory journey, understanding how the synergy of human science, contemporary digital tools, and the transformative power of AI can be harnessed to craft a society that's not just connected but also compassionate, equitable, and deeply aware!

What Is Human Science and How Do You Make It Count?

Human science, often tucked neatly within academia's hallowed halls, goes beyond mere academic exploration. It's the key to understanding the intricate ballet of human behavior, cognition, culture, and society, encompassing disciplines like psychology, sociology, anthropology, political science, and economics. It's where we decipher the age-old dance between an individual and their ever-changing societal backdrop.

In the age of the tech boom, it's tempting to assume that cold data and algorithmic logic reign supreme. But there's an inescapable truth that, regardless of how advanced our machines get, they serve a species steeped in complexity and contradiction – *us*. AI's ascension hasn't diminished the value of human science; if anything, it's amplified its significance.

Now, introduce GenAI into the mix – imagine a world where artificial intelligence isn't just replicating processes, but also reflecting, understanding, and empathizing with the human experience. Here's where the nexus of human science and GenAI becomes game-changing. A sociology algorithm that predicts not just market trends but societal shifts, an anthropological code that's sensitive to cultural nuances, or a political science AI that aids in crafting policies with a deep-rooted understanding of a populace's psyche.

For instance, consider a city's public transport system. Beyond just efficient routing, what if AI, imbued with anthropological insights, could optimize routes based on cultural events, local traditions, or socioeconomic needs? It's not about faster buses or trains, but about understanding *why* people travel.

Or, in a world rife with polarization, imagine a political science–backed AI that doesn't just analyze voting patterns, but also the undercurrents of sentiment, the silent struggles, and the unvoiced aspirations of a demographic, enabling policymakers to craft legislation that's not just reactive but visionary.

Then, there's the promise of personalized learning. Merging psychology with AI could revolutionize education. Imagine an AI tool that doesn't merely spit out test results but understands a student's unique learning style, their strengths, and their areas of struggle. Instead of a one-size-fits-all curriculum, we're looking at tailor-made learning paths for each student.

In the words of Lex Friedman, as we usher in this new era, we're not just crafting code, but "creating poetry in algorithms." As we forge ahead, we're reminded of emphasizing the "human" in every endeavor. With the combined insights of human science and the cutting-edge potential of AI and GenAI, we stand on the brink of not just technological advancement but societal evolution.

In this digital epoch, the dance between human science and technology has never been more intricate or more vital. It's not our technological prowess, but our intricate tapestry of emotions, cultures, and societal structures that make us unique. As we stand at the crossroads of a new era, the convergence of behavioral insights with technological prowess is not just an academic interest; it's a societal imperative. It is imperative to affirm the urgency of effectively conveying human science research to the masses and its subsequent translation into palpable policy. Let's look into some of them now:

1. **Refining research through technological precision**: It's not enough to just study; it's imperative that we study right. Echoing the meticulousness, our pursuit in human sciences must be grounded in rigorous methodologies. Using a blend of quantitative precision and the depth of qualitative insights ensures that our findings are both reliable and resonate with the multifaceted human experience that include historical perspectives. The traditional methods of studying human behavior and society, while foundational, can be vastly enhanced by the tools technology offers. Using artificial intelligence, researchers can analyze vast swathes of data at a scale previously unimaginable. The real magic, however, lies in combining this computational might with the nuance and depth of human understanding.

Consider how sentiment analysis, driven by AI, can glean public mood from social media trends. Yet, it is the overlay of sociocultural insights that truly contextualizes this data, making it actionable for everything from public policy to marketing strategies. Take another example, the domain of AI-driven mental health apps. Such platforms could utilize quantitative data like user interaction times or input patterns, coupled with qualitative user feedback, to create personalized mental well-being strategies. It's the difference between a generic self-help guide and a tailored mental wellness journey.

2. **Encouraging confluence through digital collaboration:** The siloed approach of yesteryears won't serve tomorrow's needs. The realms of interdisciplinary sciences such as sociology, psychology, anthropology, and even economics, when cross-pollinated, can provide groundbreaking insights. While geographic and disciplinary silos are eroding, thanks in large part to technology! Platforms now exist that foster real-time collaboration between a sociologist in London, a data scientist in San Francisco, and an economist in Mumbai. The resultant synergy can lead to groundbreaking insights that no siloed approach could ever unearth.

Furthermore, technologies like Machine Learning can then predict patterns based on this interdisciplinary data, guiding strategies in diverse sectors ranging from urban planning to healthcare. Another example to quote: AI, with its capability to process vast datasets, can be a catalyst in this collaborative endeavor. Imagine an AI tool drawing from anthropology and economics to predict the socioeconomic ramifications of a cultural shift.

3. **Translating insights into tangible change**: The culmination of human science endeavors should always be their tangible application in bettering society. Findings confined to journals serve little purpose. Close collaboration with policymakers and practitioners ensures that our deep dives into human behavior actively address societal challenges.

The vast digital infrastructure that envelopes modern society offers a unique channel to deploy insights from human science. Mobile apps, interactive platforms, and digital outreach campaigns can swiftly turn research into actionable initiatives that have a direct societal impact. For instance, an app drawing from human science research might offer personalized mental wellness strategies, leveraging user data to ensure relevance and effectiveness.

Consider an urban redevelopment project – insights
from human sciences, when amalgamated with
AI, could lead to urban spaces that aren't just
architecturally sound but resonate with the cultural
and societal nuances of its inhabitants. Spaces
where people don't just live but thrive!

As we stride forward in this age of GenAI, our compass shouldn't just
be the cold calculus of algorithms but a harmonious blend of technology
and the profound understandings of human science. The marriage of the
two promises not just advancement, but a nuanced evolution tailored for
humanity.

Looking into this further, the unfolding tableau of the 21st century,
human science plays a pivotal role in deciphering the essence of humanity.
As we stand on the cusp of a technological renaissance, the blending of
our ancestral knowledge from human sciences with the transformative
capacities of AI and GenAI provides a beacon to illuminate our path. Let's
focus on the impact of AI, GenAI, and the human psyche a bit in detail:

1. **AI – the magnifier of human patterns and
   potentials**: Let's consider the case of an urban
   city planner seeking to redesign public spaces.
   Traditional survey methods might provide
   feedback from a few hundred residents. However,
   by deploying AI that analyzes social media data,
   planners can gain insights from tens of thousands
   of real-time conversations, understanding which
   park benches people find the most serene or which
   streets are seen as unsafe at night. With AI, the city's
   pulse is not just felt; it's visualized, quantified, and
   acted upon.

2. **GenAI – bridging generations with insight**:
   Consider the marketing campaigns of a global
   beverage brand. Past campaigns might have
   resonated with millennials but perhaps miss the
   mark with Gen Z. GenAI can help by analyzing
   generational shifts in tastes, cultural nuances, and
   online interactions. An understanding garnered
   from GenAI can guide marketers to craft a campaign
   where a soda pop isn't just a drink but a symbol
   of intergenerational conversations and shared
   moments.

3. **Human psyche in the digital epoch**: The profound
   impact of our digital interactions can be seen in
   phenomena like the "Twitter mood." A recent study
   mapped global mood swings based on Twitter posts,
   revealing patterns of joy, sadness, and stress. By
   understanding these rhythms through the prism
   of human science, mental health professionals
   can better anticipate periods of societal stress or
   happiness that include the most common human
   emotions. Imagine a mental well-being app that
   sends out positive messages or coping strategies
   ahead of anticipated stress waves, all tailored to
   individual user profiles.

Yet, amid this ocean of possibility, there lurk challenges. In our quest to
harness AI and GenAI, the importance of ethics, privacy, and the sanctity
of individual experience cannot be overstated. It's not surveillance we're
after but deep-seated understanding; not cold data but warm, empathetic
insights. As we bridge the past wisdom of human sciences with today's
technological prowess, our goal is to craft a tapestry where technology
doesn't merely observe but empathizes, uplifts, and augments.

Table 3-1 highlights how these technological tools can augment various aspects of human science.

Table 3-1. *Tools Mapping Human Science*

Aspect of Human Science	Traditional Method	AI and GenAI Enhancement	Significance
Societal mood mapping	Surveys and interviews	Real-time social media mood analysis	Capture global mood patterns in almost near real-time
Generational behavior	Focus groups of different age groups	GenAI behavioral pattern recognition	Tailor products/ services to specific age demographics
Urban planning	Resident feedback forms and town halls	AI analysis of geotagged data	Create spaces resonating with real-time public needs
Marketing trends	Historical sales and trend prediction	AI prediction models	More accurate forecasts of market needs
Mental health trends	Hospital data and surveys	AI analysis of online sentiment	Proactive mental health interventions

In this table, the significance column pinpoints the distinct advantage or innovation AI and GenAI brings to the table. For instance, while traditional methods might measure mood or trends periodically, AI can provide nearly real-time insights, offering a granularity and immediacy previously out of reach. This can help sectors, whether public or private, to respond more dynamically to emerging human needs and patterns. In the later section, we shall dive a bit deeper into some of these online tools.

The digital tools at our disposal amplify the significance and reach of human sciences. Yet, it's essential to wield these tools with a sense of responsibility, cognizant of the deep cultural, emotional, and societal intricacies of the very humans these technologies aim to serve. As we propel forward, the objective isn't merely technological advancement but ensuring this progress resonates with the innate human desire for connection, understanding, and equitable growth. In this dance of bytes and beliefs, may we choreograph a future that's not just advanced, but more profoundly human.

Heading into this new era, our guiding principle must be rooted in responsibility and respect. It's more than just making human science count; it's a clarion call to ensure that in this digital age, every individual feels seen, heard, and valued.

Building Social Equity with Human Science

Social equity, often viewed through the lens of distributing resources and opportunities fairly, extends beyond mere material allocation. It delves into the nuances of understanding, interpreting, and acting upon the variances of individual and community experiences. Let's now dive right in!

Unveiling the Shadows of Inequality

Traditional perspective: Human science has, historically, been the torchbearer in revealing the multifaceted layers of social inequalities, be it glaring economic disparities, asymmetric access to education, or the chasms in healthcare availability. It's through this traditional lens that we've formed the foundations of our equity measures.

For example, studies, such as the World Inequality Report, have given us a granular understanding of wealth disparities, shaping policies like progressive taxation.

AI and GenAI augmentation: Enter AI and its evolved sibling, GenAI. These technologies don't just lay inequalities bare; they can predict emerging disparities, offering us a proactive rather than reactive approach. Yet, in this digital magnification, the risk remains of viewing individuals as mere data points.

For example, platforms like Google's AI for Social Good are being used to forecast flood events in regions like India, thus providing vulnerable populations with timely alerts.

Merit and quality: Emphasizing merit and quality means ensuring that our technological solutions don't just aggregate data, but prioritize data that genuinely contributes to understanding the human experience in its full depth.

For example, in the realm of education, Khan Academy offers quality content to millions globally, ensuring meritocratic access to information irrespective of economic backgrounds.

Data-Driven Policymaking

Traditional perspective: Policies, when formulated with empirical evidence from human science, hold the promise of tangible change.

For example, Sweden's policy to provide free higher education was informed by long-term studies which showed the correlation between education, reduced crime rates, and enhanced societal contributions.

AI and GenAI augmentation: Machine learning models can digest vast datasets to suggest which interventions may hold the most promise. However, they must be programmed with an understanding of equity, ensuring that algorithmic biases don't perpetuate existing inequalities.

For example, in the United States, the Los Angeles Fire Department has started using a machine learning model to predict and thus promptly respond to potential fires, a system that has implications for disaster management globally.

Merit and quality: Here, merit underscores the importance of prioritizing solutions that have a track record of efficacy, while quality emphasizes the value of long-term, sustainable solutions over quick fixes.

For example, Singapore's Smart Nation initiative, using real-time data to optimize everything from traffic flow to elder care, stands as a testament to quality-driven, merit-based urban planning.

Empowerment at the Grassroots

Traditional perspective: Human science isn't just about top–down examination; it's about understanding communities from within, letting those in marginalized sections voice their narratives.

For example, the Fair Trade movement empowers marginalized farmers by enabling them to share their narratives, thus influencing purchasing patterns globally.

AI and GenAI augmentation: Community-driven data collection, interpreted by sophisticated algorithms, can empower communities to track their own progress and challenges. Still, this necessitates technology that's accessible and understandable at the grassroots level.

For example, U-Report, sponsored by UNICEF, uses SMS to gather feedback from communities about relief efforts. During the Ebola crisis in Liberia, this proved instrumental in shaping timely interventions.

Merit and quality: These concepts underline the need to ensure that the most effective community-driven interventions rise to prominence and that the tools used are of a caliber that respects the users.

For example, in Kenya, the Ushahidi platform, combining local eyewitness reports with digital maps, was critical in ensuring that aid reached those in most need during post-election violence, championing both merit in data collection and quality in crisis response.

In essence, while human science offers the depth and breadth of understanding, it's through the fusion of AI and GenAI that we can scale and amplify these insights. Yet, in this alchemy of human intuition and artificial intellect, the guiding principles should be merit and quality, ensuring that we strive not just for equity and equality but for a genuinely enriched societal experience.

Current Online Tools and Apps That Indirectly Measure Your Social Behavior

In an era where our digital footprints often speak louder than our voices, understanding human behavior has transcended the confines of traditional methodologies and academic corridors. It's an age where a simple "like" on a social media post or the pattern of our online shopping cart narrates a nuanced story about our inclinations, emotions, and, sometimes, even our aspirations. The symbiotic evolution of human and machine, and the latter's quest to understand us, has birthed a plethora of online tools and applications. These tools not only mirror our societal matrix but also have the profound ability to influence and shape individual behavior. From the ways we express and communicate, to how we learn and evolve, digital platforms offer a mosaic of insights into the complexities of human behavior. It's not just about data, but the narratives hidden within those data streams.

Our behaviors, preferences, and interactions increasingly weave through a complex web of online platforms. Let's decode the significance and possibilities of these platforms while threading in concepts of AI, GenAI, merit, quality, equity, and equality.

Our every click, swipe, and keystroke adds to a vast digital tapestry of human behavior. Whether it's the music we stream, the social media posts we engage with, or the online courses we enroll in, our virtual actions are echoing in the vast chambers of big data. Companies, researchers, and forward-thinkers use these echoes to not only understand us but also to influence societal trends. Let's delve into a few more detailed examples of how technology is peeling back the layers of human psyche:

As we dive deeper, let's explore some of these revolutionary online tools and apps, dissecting their roles and potentials in painting a holistic picture of contemporary human behavior.

Social Media Analytics

Snapshot: Platforms such as *Facebook* and *Twitter* are not just about connecting with friends. Consider the 2016 US presidential election, where social media data analytics was used to understand voter inclinations and behavior. The way a user reacts to a post, the hashtags they use, and the content they share can offer a clear window into political, social, and cultural leanings. These platforms are more than just spaces for casual interactions. They're a goldmine of data reflecting societal norms, movements, and shifts.

Beyond equity: By applying AI to social media analytics, we have an opportunity to forecast societal trends. For instance, tracking a sudden surge in mental health discussions could aid in timely policy intervention.

Merit and quality: The credibility of data and its ethical use are pivotal. Filtering authentic from the inauthentic can bolster the quality of conclusions drawn.

Fitness and Wellness Apps

Snapshot: The burgeoning use of Fitbit or Headspace reflects an intertwined narrative of health and societal cues. Consider the app, *MyFitnessPal*. More than just tracking calorie intake, it offers insights into societal trends related to health, diet preferences (like veganism or keto trends), and even the psychological motivation behind food choices during certain times (say, comfort eating during lockdowns).

Beyond equity: Imagine harnessing this data to determine areas in a city experiencing higher stress levels, thus guiding urban planners or mental health services.

Merit and quality: It's not just about gathering data, but also discerning the merit of such data to catalyze tangible health outcomes.

Communication and Collaboration Tools

Snapshot: *Slack* and *Microsoft Teams*, especially in the era of remote work, offer cues about the future of workplaces. Patterns of communication, frequency, and timing of messages can provide insights into work-life balance, efficiency, and even employee well-being. These apps like Microsoft Teams, Slack, or Zoom, now staples in our professional lexicon, chronicle the evolution of workplace dynamics.

Beyond equity: Delving into collaboration patterns, we can explore how diverse teams ideate versus homogeneous ones.

Merit and quality: Ensuring that insights uphold the tenets of professional integrity and respect remains paramount.

E-learning Platforms

Snapshot: Platforms like Coursera underscore a democratization of education. Using platforms like *Coursera*, it's evident there's been a surge in AI and machine learning courses over the past few years. This indicates a workforce gearing up for a tech-heavy future, a trend further accentuated by a rise in humanities courses, signifying the importance of human science in tech-dominated spheres.

Beyond equity: Data from these platforms can guide educators about learning gaps in real time.

Merit and quality: The objective isn't just wide-reaching content, but also content that stands the tests of academic rigor and relevance.

Online Surveys

Snapshot: Tools like SurveyMonkey have revolutionized data collection. In 2020, a surge of surveys via *Google Forms* around mental health showcased society's growing awareness and emphasis on mental well-being during the pandemic. Such tools have become invaluable in gauging real-time public sentiment.

Beyond equity: By tapping into global responses, policymakers can fathom global sentiments on issues, say climate change.

Merit and quality: Crafting unbiased questions that truly capture the pulse of the public is an art and science combined.

Virtual Reality and Gaming

Snapshot: Beyond entertainment, Oculus Rift or Steam can transport us into alternate realities. On platforms like *Steam*, games that focus on world-building or societal simulations (like Sims) allow researchers to study decision-making processes, societal structures, and even economic principles in a controlled, virtual environment.

Beyond equity: Understanding behaviors in these platforms can revolutionize fields, from therapy to urban planning.

Merit and quality: The challenge? Ensuring that these virtual worlds uphold values of inclusivity and do not inadvertently perpetuate stereotypes.

Dating Apps

Snapshot: Apps like Tinder and Bumble not only revolve around romantic pursuits but also mirror societal trends in gender dynamics, cultural preferences, and even the shifting paradigms of modern relationships. These platforms are not solely about finding a romantic connection; they also serve as digital dioramas showcasing evolving societal values. By analyzing interactions, preferences, and trends on these platforms, one can gain insights into broader societal shifts. For instance, the rise in profiles mentioning terms like "feminism" or "LGBTQ+" over the years indicates a growing acceptance of diverse identities and values. When juxtaposed with AI and GenAI, these apps can potentially offer predictions about future relationship dynamics, societal acceptance of various groups, and even the evolution of gender norms. Yet, for these tools to be truly transformational and not just transactional, there's a need to integrate principles of equity, quality, and merit. This would ensure that while technology steers our interpersonal dynamics, it does so in a way that fosters understanding, acceptance, and meaningful connections.

By juxtaposing human behavior, as mirrored by these tools, against the sophisticated algorithms of AI and GenAI, we unearth a treasure trove of patterns and predictors. The digital age, bolstered by AI, doesn't just hold a mirror to society; it becomes an active participant in shaping its future. By emphasizing quality, merit, equity, and equality, there's an opportunity to leverage these insights for a harmonious and enlightened society.

By capturing the essence of human behavior through these digital tools, and overlaying it with the precision and predictability of AI and GenAI, we have an opportunity to mold a society that's not just data-driven, but also principle-led, integrating the best of human values with the power of technology.

Tying this all together, it's evident that the digital imprints we leave can be instrumental in sculpting a world that balances equality with merit, equity with quality. As we stride forward, harnessing the potential of AI and GenAI, the pivotal question remains: Will we be architects of a digital renaissance or passive bystanders? The script, it seems, is ours to write!

Asserting the criticality of communication, while delving into the technological intricacies, we can bring forth stories of real people behind these tools. The following table summarizes to capture this confluence, emphasizing not just equity and equality, but also highlighting the indomitable pillars of quality and merit. Table 3-2 delves into some real-time examples that are both informative and embody some of those nuances.

Table 3-2. *Online Tools and Apps in Human Science Context*

Tool/App	Primary Function	Equity and Equality Perspectives	Quality and Merit Implications
Social media analytics (e.g., Facebook Insights)	Gauge user behavior and preferences	Can show demographic data, ensuring diverse representation in marketing	Allows brands to refine their messaging based on high-quality user feedback
Fitness apps (e.g., Fitbit)	Monitor physical health and activity	Ensures all users, irrespective of physical capability, can set and achieve personal goals	High-quality data helps users meritocratically improve and benchmark their progress

(*continued*)

Table 3-2. (*continued*)

Tool/App	Primary Function	Equity and Equality Perspectives	Quality and Merit Implications
Communication tools (e.g., Slack)	Facilitate professional communication	Democratizes team communication, promoting equal voice irrespective of hierarchy	Enhances the quality of collaborations by tracking project merit and efficiency
E-learning platforms (e.g., Coursera)	Disseminate knowledge and skills	Promotes equal access to top-tier education globally	Allows learners to gauge course quality and select based on merit
Dating apps (e.g., Tinder)	Facilitate social connections	Diverse representation and inclusivity measures like more gender options	Algorithms promote high-quality matches based on mutual interests, ensuring a merit-based connection
VR platforms (e.g., Oculus Rift)	Immersive experience and gaming	Democratizes experiences, allowing users to virtually "live" various realities	Quality of the virtual experience combined with user feedback ensures a merit-based evolution

Let's now dive into the frameworks and methodologies behind the tools of human science a bit!

Theoretical Frameworks and Methodologies in Human Science

In the intricate fabric of human society, every strand tells a story. It's a story of patterns, behaviors, relationships, and, above all, understanding. The meticulous study of these stories, the quest to decipher the why and how of our actions and interactions, becomes pivotal. As technology, particularly AI and GenAI, enters the fray, the canvas of this understanding becomes broader, deeper, and immensely more intricate. These tools, if wielded judiciously, have the potential to not only reflect our society but to reshape it, ensuring a future where every thread, every story, matters.

Enter AI and GenAI...

Let's chime in about the transformative potential of AI and GenAI. When aligned with these frameworks and methodologies, they can refine our insights, adding layers of depth to our analyses. It's not just about delving deeper, but about the narrative we create around these findings – ensuring they resonate with the populace and effect change. Let's now pivot the discussion toward the integration of quality and merit. While equity and equality remain crucial, the true potential of these technologies lies in enhancing the quality of human science research and ensuring merit isn't sidelined.

So, let's break down these theoretical frameworks, considering the influx of AI and the new perspectives of quality and merit:

Social constructionism: Through the prism of AI, we can detect and interpret subtle patterns in our social constructs. But, we should caution, we mustn't let machines dictate these constructs but instead use them to better understand and perhaps, challenge preexisting norms. With AI's capability to interpret vast amounts of data, we can potentially unearth patterns in our societal constructs that were previously opaque.

For instance, by analyzing global news stories over decades, AI might reveal how certain narratives or stereotypes have been socially constructed and propagated.

Behaviorism: AI can meticulously track stimuli and responses over vast datasets, ensuring that not just the predominant behaviors but also the outliers are considered, ushering in a new standard of quality in behavioral studies. By using AI to track stimuli and responses across large datasets, we can gain insights into behavioral shifts on a grand scale. A real-world example could be AI analyzing consumer responses to varying marketing stimuli, helping industries adapt their strategies in real time.

Cognitive psychology: We would perhaps get animated discussing the vast potential of melding cognitive studies with AI. Beyond just examining patterns, GenAI could possibly predict behavioral shifts, ensuring preparedness in societal frameworks. The potential of integrating cognitive studies with AI is immense. For instance, using AI in conjunction with virtual reality environments, researchers can simulate decision-making scenarios and predict behavioral outcomes based on cognitive processes.

Ethnography: In an era where digital ethnography is gaining traction, AI can assist researchers in deciphering large sets of qualitative data, ensuring the essence of cultures is captured with nuance and depth. Digital ethnography, backed by AI, can assist researchers in sifting through and interpreting vast sets of qualitative data. Platforms like YouTube, for instance, are goldmines for digital ethnographers, where user-uploaded content can provide authentic insights into cultural behaviors, practices, and values from around the world.

Social network analysis: The complex webs of our social networks hold keys to societal patterns, trends, and shifts. We might underline the fact that quality insights from such analyses, backed by AI's computational prowess, could redefine how we approach social dynamics. Social networks are intricate mazes of human interactions. AI can help decipher these, offering insights into societal patterns and shifts. An example here would be analyzing Twitter data to track the spread and influence of specific ideologies or trends over time.

In essence, while theoretical frameworks offer a lens, AI and GenAI can be the instruments that refine, focus, and amplify our view. The goal isn't just equitable access to insights but ensuring these insights uphold the highest standards of quality and merit. While our traditional frameworks offer a lens to view the human story, AI and GenAI sharpen and enhance this lens. They don't just give us clearer vision – they open our eyes to aspects we were previously blind to. The promise lies not just in equitable access to insights but in ensuring these insights uphold the highest standards of quality, merit, and humanity.

Let's now briefly touch base applications of human science while addressing many of these challenges across the globe!

Applications of Human Science While Addressing Global Challenges

In an era where technology's tendrils touch every aspect of our lives, the interplay of human science with tools like AI and GenAI promises a nuanced approach to our world's multifaceted challenges. The potential of these technological marvels isn't just in their computational prowess, but in their ability to enhance our understanding of the delicate human threads that weave the global tapestry. However, while the allure of equity and equality in technology is evident, we mustn't overlook the nuanced power of quality and merit, ensuring our solutions are not just well-intended but also exceptional in their efficiency and effectiveness.

In this digital age, entwined with a post–COVID-19 world's challenges, the blending of human science with the powerful capabilities of AI and GenAI is both a necessity and an opportunity. Beyond just equity and equality, the pivot towards integrating quality and merit is essential to craft solutions that are not only inclusive but also effective and exceptional in addressing societal challenges. The post-pandemic world has underscored the significance of this harmonization, as evident in some real-time examples that we shall discuss in the following.

Climate change and environmental sustainability: While science provides data on changing weather patterns, human science digs deep into the collective psyche, assessing why certain populations resist sustainable practices. Integrating AI could predict which cultural shifts might encourage eco-friendly behaviors. For example, AI-backed campaigns in areas with historical resistance to sustainable practices could leverage local cultural nuances, increasing their effectiveness.

The lockdowns during the pandemic resulted in cleaner air and waters, offering a glimpse into a possible sustainable future. However, while some celebrated this short reprieve, others faced job losses from industries deemed "unfriendly" to the environment. Human science helps understand these sociocultural dynamics. For instance, using AI, governments could identify regions most affected by such job losses and initiate green skill development programs, ensuring that the transition to sustainability is not just equitable but also of high quality and beneficial in the long run.

Poverty and inequality: Behind stark numbers and statistics, there are stories, traditions, and entrenched societal norms. Human science deciphers these, and with the precision of GenAI, we can tailor interventions to specific communities. A practical application could be AI analyzing trends in local job markets and suggesting skill development programs specific to regions, ensuring both equality of opportunity and quality of outcome based on individual interests.

The pandemic exacerbated socioeconomic disparities. While some businesses flourished in the digital wave, traditional sectors, especially in regions with limited technological adaptability, suffered. Through human science, one can uncover the layers of these disparities. GenAI could, for example, assist in crafting hyperlocal economic policies. By recognizing the merits and strengths of individual regions, policies can be more tailored, offering both equal opportunities and ensuring that the solutions provided are of superior quality.

Mental health and well-being: As the world grapples with a mental health crisis, understanding cultural perceptions and societal stressors becomes imperative. AI, when integrated with platforms like social media, could offer insights into collective mental well-being, highlighting potential areas of concern. For instance, a region with a significant uptick in searches related to mental health challenges could be provided with targeted resources, ensuring timely interventions.

Isolation during the pandemic led to a global uptick in mental health challenges. Yet, the acceptance and approach to mental health vary culturally. Human science can pinpoint cultural hesitancies or barriers. AI could be used to design culturally sensitive awareness campaigns, ensuring quality mental health information reaches even the most remote or resistant areas. An example would be using AI to analyze social media trends in specific regions, offering interventions like virtual therapy or counseling when a potential mental health crisis is detected.

Conflict and peace-building: The roots of conflict often lie in historical, cultural, and social nuances. Human science attempts to untangle these. GenAI could run simulations based on historical data to predict possible conflict areas and aid in formulating proactive peace-building strategies. An example might involve AI-powered dialogue platforms that bridge communication gaps between conflicting factions, facilitating understanding and reconciliation.

The post–COVID-19 era saw a rise in conflicts, from debates over vaccine distribution to geopolitical tensions over border closures. Understanding the undercurrents of these conflicts requires an in-depth exploration of societal norms, historical grievances, and cultural perspectives. AI and GenAI could play pivotal roles here. For instance,

AI-powered platforms might mediate dialogues between nations over vaccine distribution, ensuring discussions are not just equitable but prioritize the merit and quality of arguments presented.

Harnessing the best of human science and the precision of AI and GenAI ensures that as we strive for a more equitable world, we do not compromise on the quality of our solutions. Equity, equality, quality, and merit need not be at odds – instead, they can form the pillars of a harmonious, forward-thinking global society.

The post–COVID-19 world is complex, with multifaceted challenges that require nuanced solutions. By anchoring our strategies in human science and augmenting them with the precision of AI and GenAI, we can ensure that our path forward cherishes not just equity and equality, but also the inherent value of quality and merit.

Conclusion

Our lived experiences, rife with the delicate dance of societal norms, behaviors, and interactions, paint a rich tableau. Human science seeks to decode this tableau, offering profound insights into the forces that mold human behavior and the fabric of our societies. The aim isn't just to understand but to apply this knowledge to the world's most pressing issues, forging a path that respects both the individual and the collective.

In our technologically saturated world, tools like AI and GenAI stand out not merely as novelties but as transformative forces. Yet, while these technologies can analyze patterns and predict outcomes with an accuracy that once seemed the stuff of science fiction, they're most effective when complemented by the depth of human understanding. It's like having a ship (AI and GenAI) that can navigate any waters, but human science provides the star by which to chart its course.

Quality and merit are the unsung chords in the symphony of progress. In our rightful quest for equity and equality, we must not overlook the profound significance of quality, that is, the degree of excellence, and merit, that is, the inherent value or worth of something. For example, while AI can ensure that educational content reaches every nook and corner of the globe, human science ensures that this content is culturally relevant and meaningful. GenAI can tailor health solutions to individual genetics, but it's the human-centric approach that ensures these solutions resonate with the individual's sociocultural milieu.

Drawing this journey to its nexus, the blend of human science with AI and GenAI is akin to a maestro orchestrating a harmonious blend of classical and modern instruments. As we stand at the crossroads of tradition and innovation, the post–COVID-19 epoch beckons us to be both visionary and rooted. It implores us to champion equity and equality while simultaneously upholding the torches of merit and quality. In doing so, we'll not only address the challenges of today but also lay the foundation for a future that's equitable, sustainable, and truly resonant with the human spirit.

As we navigate the intricate tapestry of human behavior, there's a palpable realization that our modern era is defined not just by data, but by the principles that guide its use. We stand on the precipice of a world where online tools and applications have transformed not just how we interact, but fundamentally how we understand ourselves. Equity and equality form the bedrock of these digital interactions, ensuring that every voice, irrespective of background or standing, has a stage. Yet, it would be myopic to stop here. The stories that these tools unravel emphasize the paramount importance of quality – the finesse of the tool, the depth of the insight. And then, there's merit – the hard-earned results, the feedback loops, and the pure unadulterated drive to be better. This is the new age renaissance of human science, a confluence of old principles and new technologies.

But what's the next frontier? If online tools and apps are the vessels, then networks are the vast oceans they navigate. The interconnectedness of our age is not just physical but deeply digital. In the upcoming chapter, we dive deep into the world of networks – understanding their rise, their influence, and their undeniable impact on human science and behavior. From the intricate algorithms that dictate our social media feeds to the sprawling global connections that influence economies, networks are reshaping the very fabric of our society. As we journey through this vast digital landscape, we'll explore the beautiful balance between individual nodes and the grander network, between the singular and the collective, and between the micro and the macro.

CHAPTER 4

Networks and Their Impacts

From the intricate webs of our global economy to the vast expanse of the digital realm, networks represent the backbone of our interconnected age. They manifest as structures of relationships, binding diverse nodes in an intricate dance of interdependence. As we embark on this exploration, one quickly realizes that it isn't just about the connections, but the quality of these connections and the merit they bring to the table. While the world today largely champions the principles of equity and equality, ensuring that all nodes in a network, regardless of their position or prominence, have a say, there's a delicate balance to be struck. And that balance revolves around the notions of quality – the substance behind these connections – and merit – the value and innovation they add to the larger ecosystem. By understanding and appreciating this balance, we position ourselves to harness the true potential of networks in our lives.

The Science Behind Networks

To truly fathom the power and reach of networks, one must begin at their foundation. Every network, whether digital or physical, is made up of nodes and links. Think of it as the neurons in our brain, constantly firing and creating pathways of thoughts and memories. However, the true potency of a network isn't merely in its size but in the strength and

© Raghu Banda 2024
R. Banda, *Building Social Equity with AI*, https://doi.org/10.1007/979-8-8688-0091-7_4

efficiency of its connections that add value within the network. Modern studies show that networks operate on the principle of "six degrees of separation," suggesting that any two people on Earth can be connected in as few as six intermediary relations. In such a dense, interconnected world, the efficiency and quality of these connections become paramount.

The Social Implications of Networks

Beyond science, networks play a pivotal role in the social dynamics of our era. Platforms like Facebook, Twitter/X, TikTok, and LinkedIn are more than just technological marvels; they are socioeconomic platforms that dictate cultural trends, drive business opportunities, and, sometimes, even shape global politics. But as networks grow and evolve, they inherently come with challenges of equity and equality. While the digital age has democratized information, it has also created echo chambers, highlighting the importance of diverse connections and the need for a merit-based algorithm that rewards genuine and relevant contributions.

Strategies for Optimizing Networks

To leverage the vast potential of networks while minimizing their pitfalls, we must adopt holistic strategies. Emphasizing the quality of connections ensures that networks remain robust and resistant to misinformation. Championing meritocracy within these platforms encourages innovation and genuine contributions. Furthermore, by ensuring equity and equality, we foster a sense of inclusivity, ensuring every node, irrespective of its origin or prominence, contributes to the collective growth.

How Networks Impact Humans in Transactions and Interactions

In our quest to elucidate the extensive impacts of networks, we must embark upon a meticulous deconstruction of their intrinsic properties, the silent symphonies they orchestrate between their constituents, and the harmonious yet often complex dialogues that unfold within their parameters. Networks, in their most rudimentary form, are akin to an elaborate fabric, weaving together a myriad of nodes – be they individuals, institutions, or inanimate interfaces – through an array of relationships delineated by the invisible threads of connections.

Deconstructing the Anatomy of Networks

Every person, every organization, every computational interface that we encounter, articulates a unique melody in the grand orchestra of our interconnected reality. In the world of social media, platforms like Facebook epitomize the ethos of networked existence, showcasing a universe where every individual is not just a passive recipient of information but an active node, constantly transmitting and receiving data. Here, we see equity and equality in their most vibrant manifestation, a world unhindered by the physical barriers of yesteryears.

The Multifaceted Nature of Networks

Yet, networks extend far beyond the digital corridors of social media. In the enigmatic lanes of our neural pathways, where billions of neurons engage in an intricate dance, echoing the symphony of our thoughts and emotions, there lies another example of networks. Each neuron, like a diligent artist, contributes to the masterpiece of our cognitive faculties. However, this isn't a plain field of equal contribution; it's an arena where merit and quality stand as the silent arbiters, dictating the influence of each node.

AI As the Catalyst of Evolution

In this intricate milieu, artificial intelligence (AI) emerges not just as an observer but as a dynamic participant. AI observes, learns, and evolves, absorbing the nuanced choreographies of these networks and amplifying their capabilities. Whether it's enhancing the connectivity within social networks, ensuring that the silent voices are heard, or augmenting the precision of neural networks, AI stands as the bridge connecting potential with realization.

Here, the convergence of equity and quality, equality and merit is not just desirable but essential. AI, with its impartial algorithms, ensures that every voice within a network is accorded its due space, validating the principle of equity. Concurrently, it underscores merit, ensuring that quality isn't drowned in the cacophony of voices but is amplified, contributing to the collective wisdom.

Types of Networks

In the expansive terrain of our interconnected existence, networks, akin to the silent architects of our social, informational, and biological ecosystems, craft the pathways that facilitate the seamless exchange of value, ideas, and life itself. A network isn't a static entity but a dynamic, evolving structure, adaptive and responsive, echoing the intricate rhythms of its diverse constituents. Each type of network, with its distinct characteristics, weaves a unique narrative of connections, becoming a living testament to the intricate symphony of interactions that define our world.

In an era where the invisible threads of connectivity weave through every aspect of our existence, networks have emerged not just as structural frameworks but as evolving ecosystems pulsating with dynamic energy, fostering relationships, facilitating information flow, and echoing the intricate dance of biological life. Here, each form of network, distinct yet interconnected, unfolds a narrative that is as profound in its simplicity as it is complex in its multifaceted interactions.

- **Unraveling the complexity of social networks**: Consider the mosaic of social networks – a dynamic tableau where human connections, emotions, and exchanges find a virtual playground. Facebook, Twitter/X, LinkedIn – these aren't just platforms; they are microcosms reflecting the sophisticated elements of human interactions. Equity and equality echo in these spaces, painting a canvas where every individual, every voice, is acknowledged. Yet, amid this harmonious inclusivity, the essence of quality and merit is not just preserved but celebrated. Each post, each connection, is a testament to the enriching diversity that fosters innovation and creativity.

  Example: LinkedIn's algorithm, underpinned by AI, ensures not just that diverse professionals connect but that the quality of these connections is enhanced, fostering collaborations that are rooted in mutual respect, innovation, and expertise.

- **Navigating the information networks**: The realm of information networks is akin to an expansive ocean of knowledge. The World Wide Web, citation networks – they are repositories where data and information flow unbridled. In this universe, AI emerges as the navigator, discerning, filtering, and amplifying content that is not just relevant but valuable, ensuring that in the pursuit of equality and accessibility, the sanctity of quality and merit is unfettered.

  Example: The Google Search algorithm, enriched by AI, tailors search results, ensuring that users access information that is not just abundant but accurate, reliable, valuable and relevant, weaving a narrative of informational equity enriched by quality.

- **The odyssey of transportation networks**: In transportation networks, the harmony of connectivity and accessibility sings the ballad of equity. Road systems, airline routes, railway networks – they are the veins and arteries ensuring that connectivity is not a privilege but a universal testament to human ingenuity. Yet, in this narrative, the undercurrents of quality and efficiency are omnipresent, underscored by meticulous design, strategic planning, and technological innovation.

  Example: Tesla's navigation system, enriched by AI, not only ensures optimal routes for drivers but underscores the essence of quality, efficiency, and safety, embodying a harmony where accessibility and quality are not just balanced but are mutual companions.

- **The enigma of biological networks**: Biological networks, from the intricate neural networks in our brains to the mesmerizing protein interaction networks in our cells, are the sere tunes of life's path. Here, the journey is not just of discovery but of profound reverence, where AI becomes the silent observer, the meticulous documenter, and the innovative enhancer.

  Example: DeepMind's AlphaFold's groundbreaking strides in understanding protein folding is an ode to this symphony, where biological mysteries are unraveled with a precision and accuracy that echo the harmonious dance of equity, quality, and merit.

Impacts of Networks on Human Transactions and Interactions

The silent, omnipresent entities we recognize as networks are not just structural constructs; they manifest as dynamic ecosystems influencing every aspect of human life. Their influence extends from the depths of individual human interactions to the vast expanses of global economies. In this intricate dance, each note and rhythm is a testament to the harmonious coexistence of equity, equality, quality, and merit.

The Symphony of Communication

Communication isn't just an exchange of words but a communion of ideas, emotions, and innovations. In this space, networks are the invisible threads weaving through diverse landscapes of human existence. Consider the versatility of platforms like WhatsApp and Zoom. They are not mere tools; they embody a universe where communication is democratized, yet the richness of content, the depth of dialogue is exalted.

Example: Zoom, during the COVID-19 pandemic, emerged not just as a tool for virtual meetings, but as a sanctuary where ideas flowed freely, collaborations were born, and the essence of quality dialogue was celebrated amid an environment of universal accessibility. For example, AI-driven features like automated transcription, background noise reduction, and virtual backgrounds are not just technical enhancements; they are testimonies to a world where communication is adorned with the elegance of quality.

Decisions at the Crossroads of Information Flow

In the world of decision-making, networks are the silent guides, illuminating paths, and shaping choices. They wield their influence not with authority but with the silent eloquence of information and insights. Here, algorithms, like those embedded within Google Search or Amazon's recommendation systems, echo the nuanced balance of personalization and universal access.

Example: Amazon's recommendation algorithms, backed by AI, personalized shopping experiences, showcasing a world where choices are shaped not just by accessibility but enriched by the personalized touch of quality and relevance. Amazon's recommendations are precision personified, thanks to AI. Every suggestion is a blend of universal access (equality) and personalized relevance (quality), ensuring that choices are not just abundant but meaningful.

A Dance of Relationships in Social Connections

Social relationships, like the finely woven threads of fabric, are shaped and enriched by networks. Platforms like Facebook, TikTok, YouTube shorts, and Instagram are canvases where relationships paint their narratives. Each connection, each interaction, echoes the silent symphony of equity and equality, where diversity is celebrated, and the nuances of individual uniqueness are revered.

Example: Instagram, with its global reach, weaves a narrative where diverse cultures, ideologies, and voices find a universal canvas, yet the uniqueness of individual expression, quality of content is celebrated with exuberance. Instagram's AI-powered algorithms ensure that every user's feed is a personalized narrative, echoing the silent symphony of universal accessibility enriched by the personalized touch of quality.

The Alchemy of Economic Prosperity

Economic opportunities, in the embrace of networks, transform from isolated events to global phenomena. E-commerce platforms, digital currencies, and online marketplaces are landscapes where opportunities are not bound by geographies or privileges. Here, the dance of equity and quality is most profound, echoing in the seamless transactions on platforms like eBay or the decentralized finance ecosystems emerging from blockchain technology.

Example: eBay's global marketplace ensures that sellers, irrespective of their geographical location or scale, have a platform to reach global audiences. Yet, amid this universality, the platform's rating and review system ensures the quality of transactions is integral. AI in eBay is not just about connecting buyers and sellers; it's about ensuring that every transaction, every exchange is characterized by a seamless blend of accessibility (equity) and excellence (quality and merit).

As we weave through these complex terrains, AI emerges as the silent yet potent force, ensuring that the dance of networks is not just about connections but about meaningful, enriched, and valued interactions. It's a world where the pragmatism of connections meets the poetry of quality, where the universal access of equality waltzes elegantly with the personalized touch of merit.

The Science Behind Networks and Their Social Impact

The synergy of mathematics, computer science, and sociology in network science unveils a canvas where structures, behaviors, and innovations flourish. Here, each strand of connection, every node, is a testament to the complex cadence of complexity and simplicity, individuality and universality. The complexity of network topologies, robustness, and centrality unveils a universe where equity, equality, quality, and merit are not disparate stars, but constellations that illuminate the landscapes of human interaction and innovation.

Network Topology: A Melange of Diversity and Complexity

Consider the intricate world of social media platforms like Twitter a.k.a. X. Every tweet, hashtag, and connection unfolds in a space where network topology is as diverse as the global populace it represents. It's a world where every node is a voice, and every edge is a conversation. Here, the robustness isn't just technical; it's profoundly social, echoing the resilience of human connections in the face of disruptions.

Example – Twitter a.k.a. X: AI steers this dance of conversations. In the rhapsody of trending hashtags, AI ensures that the chorus of universal participation (equality) is adorned with the personalized touch of individual relevance (quality and merit).

Network Robustness: The Resilience of Connections

In the realms of transportation, let's traverse the terrains of airline networks. Every route is a testament to the science of robustness, a narrative where technical precision meets social resilience. Amid the rhythms of arrivals and departures, there lies the silent yet potent narrative of human mobility, economic prosperity, and social connections.

Example – airline networks: AI, like an adept conductor, orchestrates this complex symphony. Every flight, every route, is a narrative where safety (quality) and accessibility (equity) coalesce, ensuring that the journey and experience is as significant as the destination.

Network Centrality: The Epicenter of Influence and Innovation

LinkedIn, the professional networking behemoth, is a canvas where the theory of network centrality unfolds in real time. Every connection, endorsement, and post is a strand in this intricate web where professional identities are not just showcased but nurtured and enriched.

Example – LinkedIn: AI in LinkedIn is not just an algorithm; it's a mentor, a guide that ensures that every professional's journey is characterized by the universal access to opportunities (equality) while being enriched by the personalized pathways of career growth (quality and merit).

In this world of networks, the rigid boundaries of disciplines dissolve, unveiling an interdisciplinary narrative that's as profound as it is pragmatic. Every strand of mathematics, echo of physics, innovation of computer science, and insight of sociology, is a chapter in this unfolding narrative of networks.

Network science isn't just a theoretical construct but a lived reality, a silent yet potent force that shapes our choices, behaviors, and innovations. In the subsequent pages, we shall immerse deeper, exploring the facets of this dynamic field where the digital pulse of AI meets the organic rhythms of human behaviors, where the universal narratives of equality and equity dance elegantly with the individual threads of quality and merit, weaving a tapestry of connections that's as intricate as it is profound.

Social Networks and Their Impacts

Navigating the intricate pathways of social networks, we encounter a dynamic ecosystem where information, influence, and innovation pulsate with vibrancy. Each connection, like threads woven into a tapestry, illustrates the power of collective influence, shaping societal norms, cultures, and attitudes. It's in this intricate dance of connections that the threads of equity and equality intertwine with the melodies of quality and merit, echoing the symphony of a socially harmonious world.

Information Diffusion: The Ripple Effect

Dive into the realm of platforms like Twitter or Reddit. Here, a tweet or a post isn't merely a string of text but a pebble that, when dropped into the waters of public discourse, creates ripples of influence. Information, ideas, and perspectives disperse, impacting decisions and opinions, reflecting the pulsating energy of democratic dialogue.

Example – the Reddit effect*:* In the arena of AI, we see algorithms playing the role of silent architects, shaping the pathways through which information travels. Through machine learning and natural language processing, AI ensures that information diffusion is not just rapid but is nuanced, echoing the complex interplay of universal access (equality) and contextual relevance (quality and merit).

Social Capital: The Currency of Connectivity

On platforms like LinkedIn, connections aren't merely digital links but conduits of social capital. Each endorsement, message, or shared post is an investment in a social economy where the dividends are trust, cooperation, and enhanced individual capacity.

Example – LinkedIn AI: Here, AI is also the custodian of social capital! By personalizing connections and content, AI ensures that social capital is not merely accumulated but is strategically invested, echoing the balanced dance of universal access (equity) and individualized value (quality and merit).

Social Influence and Conformity: The Double-Edged Sword

Take Instagram, where trends are born and opinions are shaped. Here, every like, comment, or share is a nod of social approval, a subtle yet powerful force that molds attitudes and behaviors while influencing one's decision-making.

Example – Instagram's algorithm: AI, in this space, is both a mirror and a window. It reflects the prevailing social norms (conformity) while opening windows to diverse perspectives (quality), ensuring that social influence is a catalyst of cultural richness, not just uniformity.

Social Segregation and Homophily: The Echo Chambers

Facebook's news feeds are a narrative of affinity, where likes and shares carve out spaces of similarity, echo chambers where voices resonate with concordant tones.

Example –Facebook's filters: AI here is a bridge builder. While echo chambers resonate with the harmony of similarity (equality), AI algorithms introduce dissonant yet enriching notes (quality and merit), weaving a tapestry of social discourse that's vibrant with diversity.

Strategizing Network Enhancement for Optimized Outcomes

As we harness AI's potential to optimize network outcomes, the focus intensifies on designing algorithms that are not just intelligent but are also embedded with ethical and social sensibilities. AI in this landscape isn't just a tool; it's a partner in the intricate dance of social dynamics, where the rhythms of equality and equity are adorned with the melodies of quality and merit.

Our journey into the world of networks, where connections are as fluid as they are firm, where influences are as subtle as they are significant, doesn't conclude here. As we pivot to the subsequent narratives, we delve deeper into the ecosystems of networks. A world where the mathematical precision of algorithms and the organic spontaneity of human interactions coalesce, unveiling landscapes where the science of networks meets the art of human connections, echoing the symphony of a world harmonized in equity, pulsating with quality, and radiant with merit.

Improving Networks for Better Process Outcomes

In the multifaceted realm of networks, the art of connection is a jive between the tangible and intangible, between the seen corridors of connections and the unseen alchemy of interactions. The efficiency and efficacy of these networks aren't merely the summation of their structural architecture but are significantly amplified by the depth, quality, and fluidity of connections.

Building New Connections: The Genesis of Synergy

The act of building connections goes beyond the mechanistic approach of creating links. It's an art where mutual aspirations meet, igniting the genesis of collaborations that are both potent and transformative.

Example – LinkedIn: Here's a stage where professional identities are not just profiles but nodes in a global network. When AI steps into this arena, it's not just about connecting dots anymore but about identifying potential symphonies of collaborations. Every recommendation, every connection request is filtered through the lens of AI, ensuring that equality and equity are not lost words but are manifested in every interaction, while quality and merit are the intrinsic metrics that guide these connections.

Strengthening Existing Connections: Deepening the Roots

The reinforcement of existing connections is akin to nurturing a tree, where the strength isn't just in the branches that reach out but in the roots that anchor. It's a process of nurturing relationships with nutrients of trust, information, and cooperation.

Example – Slack: In platforms like Slack, conversations aren't linear threads but are woven fabrics of organizational culture. AI here is not an observer but a participant, a catalyst that ensures that every interaction is enriched with insights, every communication fosters a culture where equity is the norm, quality the expectation, and innovation the outcome.

Optimizing Network Topology: The Symphony of Efficiency

An efficient network is a cosmos where paths are not rigid trajectories but fluid avenues adapting to the pulsating rhythms of information and interactions. It's a landscape where the length of the path is not a measure of distance but the richness of exchanges.

Example – Google's BERT: In the world of information networks, AI algorithms like Google's BERT are not just about retrieving information but about understanding the nuances, the underlying desires, and the unspoken needs. Every search result is a reflection of an ecosystem where equity in information access, quality of content, and merit of sources are the harmonious notes that create the melody of information exchange.

In the heart of this dynamic swirl of connections, there's a silent yet potent force – the ethos of equity and equality, quality and merit, all interwoven intricately, each a thread in this rich tapestry of networks. As AI and human insight amalgamate, a new epoch of networks is envisioned – where connections are not just counted but valued, not just built but nurtured, and where the essence of human spirit is the melody that orchestrates this grand symphony of interconnectedness.

As we take this exploration a step further in this chapter, we will journey into an era where networks aren't just structural entities but living ecosystems. Every node, every connection is infused with consciousness, and every interaction is a step toward a world where technology and humanity converge, unfold, and transform. Welcome to a narrative where networks are the canvases, and human insights, AI, quality, and equity are the colors that bring these canvases to life.

Promoting Network Diversity

In the kaleidoscope of networks, a spectrum of diverse hues of thoughts, innovations, and creativity emerges, not just as a byproduct of varied connections, but as the cornerstone of networks thriving in adaptive resilience. The canvas of network diversity isn't monochromatic; rather, it's a multifaceted mural, each shade and tone reflecting the rich amalgamation of diversity – not just of disciplines, but of thoughts, perspectives, and ideation.

Encouraging Cross-disciplinary Collaboration: The Alchemy of Ideas

In the panorama of cross-disciplinary collaborations, two seemingly distant stars collide, creating a galaxy of novel constellations of ideas and innovations.

Example – interdisciplinary AI applications: Consider the intersection of AI with healthcare and environmental science. It's not merely about machine learning algorithms diagnosing diseases or predicting climate patterns. Instead, it's a harmonious symphony where

AI, grounded in principles of equality and equity, ensures that the diagnostic algorithms are unbiased, the predictive models equitable, and innovation is not the privilege of a few but a shared legacy of humanity. Every prediction, every diagnosis is tinted with the shades of quality and merit, ensuring excellence isn't a serendipity but a norm.

Fostering Inclusivity: A Spectrum of Perspectives

Inclusivity isn't a passive act of opening doors; it's an active endeavor of inviting, embracing, and celebrating the mosaic of perspectives and insights.

Example – open source AI platforms: Platforms like GitHub aren't just repositories of code but are living ecosystems of collaborative ingenuity. Every contributor, regardless of their geographical or social coordinates, is a node, and the collective intelligence is a network woven with threads of equity, each contribution evaluated on the scales of quality and merit, painting a narrative where diversity is the brush, and innovation the artwork.

Facilitating Bridging Ties: Weaving the Threads of Unity

The narrative of bridging ties isn't scripted in the language of connectivity alone but is narrated in the eloquent expressions of shared insights, resources, and collective aspirations.

Example – global AI initiatives: Consider initiatives that unite AI talents globally to combat challenges like the COVID-19 pandemic. The network isn't defined by the number of participants but by the richness of collaboration. Each connection is a conduit where information, insights, and innovations flow freely, unbounded by territorial or institutional barriers, ensuring that the solutions crafted are imbibed with principles of equality and equity, and excellence is the shared horizon we collectively voyage toward.

Here, diversity isn't an element; it's the soul that breathes life into connections, ensuring that every interaction is a melody where notes of equity, quality, and merit create harmonies of innovations that are both inclusive and exemplary. It paints a narrative where networks aren't just conduits of connections but are living canvasses where every stroke of innovation, every hue of creativity is a testament to the collective journey of a humanity that's diverse yet unified, distinct yet connected!

Enhancing Network Robustness

As we pivot into the realms of network robustness, we meander through the intricate pathways where networks aren't just threads of connections but are living entities, pulsating, adapting, and evolving. The narrative of robustness isn't sculpted in the rigid frames of structural fortitude but is a dynamic ballet where adaptability, redundancy, and monitoring are the choreographers crafting a boogie resilient in the face of tumults and graceful in adaptability.

Implementing Redundancy: The Multiplicity of Pathways

Redundancy isn't an architectural afterthought but a premeditated design where each alternative path is a testament to the network's intrinsic resilience.

Example – Internet infrastructure: Consider the Internet, a quintessential model, pulsating globally, yet resilient. When a natural disaster impairs a submarine cable, the Internet doesn't stutter; it gracefully pirouettes through alternative pathways. It's not just about equality of access but the merit of quality, ensuring that every byte of data flows with the assurance of uninterrupted accessibility and the fidelity of quality.

Monitoring Network Health: The Pulse of Functional Grace

Monitoring isn't a passive act of observation but an active engagement where insights gleaned are arteries feeding the adaptive evolution of networks.

Example – healthcare information networks: In the world post–COVID-19, healthcare information networks aren't just conduits of data but are lifelines. Every piece of real-time data is a pulse, and AI is the vigilant sentinel ensuring that the pulse is robust, the data flow unimpeded, and the insights gleaned are equitable in accessibility and exemplary in quality.

Developing Adaptive Capacity: The Ballet of Evolutionary Grace

Adaptive capacity isn't a feature; it's the inherent characteristic of networks sculpted for evolutionary resilience.

Example – adaptive AI algorithms in finance: In the fast-paced alleys of financial markets where volatility is a constant companion, AI algorithms aren't rigid structures but adaptive entities. Each market fluctuation is a note, and AI algorithms step to the tunes, not just adapting but evolving, ensuring that financial insights and forecasts are not just equitable in accessibility but are steeped in the merits of quality and precision.

As we unfurl this web of connections further, each thread woven with the precision of AI and the grace of human ingenuity, we aren't just spectators but active participants. We are in a world where networks aren't defined by structural robustness alone but are characterized by their adaptive grace, where every connection is a dialogue, and every node, irrespective of its geographical or digital coordinates, is a participant in this global ballet.

Each connection, each node, each pathway isn't just a structural entity but is a living testament to a world where the principles of equity and equality aren't just architectural foundations but are the soulful melodies that define to the tune of networks, where quality and merit aren't architectural afterthoughts but are the choreographers crafting a narrative of resilience, adaptability, and evolutionary grace.

Leveraging Network Analytics

Network analytics sits at the confluence of innovation and insights, a dynamic entity, constantly evolving and morphing, offering a gateway to not just observe but to comprehend and predict the intricate dance of connections and nodes. It is not merely a tool but an oracle, delineating the intricate contours of connectivity, transcending the rudimentary metrics of structure and unveiling the multifaceted dimensions of influence, community, and evolution.

Identifying Key Actors: The Anchors of Influence

In the universe of connections, not all nodes are created equal. Some emerge as the epicenters of influence, not by the virtue of volume but the gravity of quality.

Example – Twitter/X influencers: The universe of Twitter/X isn't defined by the constellation of tweets but the magnetic pull of influencers. Here, AI, with its predictive analytics, sifts through the noise and adorns those not just echoing the volumes of followers but are the harbingers of quality content. It's an intricate jive where equity in voice meets the merit of influence.

Detecting Communities: The Galaxies of Cohesion

Communities aren't defined by the proximity of connections but the intensity of interactions, where the echoes of shared beliefs and values construct the harmonious ballet.

Example – LinkedIn Professional Groups: In the professional sanctuaries of LinkedIn, AI identifies not just clusters but harmonious ensembles where professionals, though diverse, echo the symphony of shared expertise and insights. It's a narrative where the ethos of equality is in a melodic synchrony with the sonnets of meritocracy.

Analyzing Network Evolution: The Saga of Transformation

Networks aren't static monuments but dynamic entities, echoing the rhythmic ballet of evolution, a narrative sculpted not in isolation but the dynamic dialogues of connections.

Example – evolution of research networks: Consider the sanctified realms of academic research. Here, the narrative isn't static. AI, with the precision of analytics and the grace of insights, maps the evolutionary trajectory. Every research paper, every citation is a pulsating node, and AI unveils the narrative, where the symphony of equality in participation waltzes to the tunes of merit in contributions.

As we embark on this odyssey, networks unveil themselves as living entities, not defined by the rigidity of connections but the fluidity of interactions, where AI is not an observer but a choreographer, crafting a narrative where the echoes of equality and equity are in a harmonious dance with the rhythms of quality and merit.

Case Studies and Examples of Networks and their Impacts

Let's briefly discuss some of the case studies to explain networks and their impacts.

Case Study 1: The Impact of Social Networks on Job Search

Navigating through the intricate labyrinth of job seeking is an expedition, often solitary, but invariably influenced by the unseen yet palpable forces of networks. It's a sphere where the apparent asymmetry of connections unfolds a narrative, profound and insightful, echoing the harmonious interplay of quantitative metrics and qualitative insights, painted gracefully by the meticulous strokes of AI.

A Symphony of Ties: The Unseen Ensemble

The narrative of job seeking is a resonant symphony of "ties" – an ensemble where the conspicuous robustness of "strong ties" is often overshadowed by the silent yet potent resonance of "weak ties."

Real-time example – LinkedIn connections: Consider the nuanced dynamics of LinkedIn connections. AI, wielding the baton of analytics, orchestrates a narrative where "connections" transcend the numerical metrics and echo the qualitative harmonies of engagement, influence, and opportunity.

The Artful Dance of Metrics: A Canvas Painted by AI

Table 4-1 is not a static portrait of numbers but a dynamic canvas, where every metric is a pulsating note, echoing a narrative painted by the meticulous strokes of AI.

Table 4-1. *Job Seeker Success Rates Based on Network Characteristics*

Network Characteristic	Success Rate (%)
Strong ties only (close friends or family)	17
Weak ties only (acquaintances)	56
Mixed ties	27

Interpretation and insight – the reflections of equity and merit: In the tableau of job seeking, equity resonates in the universal access to opportunities while merit orchestrates the personalized narratives of skills, competencies, and aspirations. Here, the 56% success rate of "weak ties" isn't an isolated metric but a complicated choreography of accessibility, diversity, and opportunity.

The Unfolding Narrative: A Journey Beyond Numbers

As we delve deeper, this isn't a static illustration but a dynamic narrative, where the geometric lines and curves echo the silent melodies of aspirations, opportunities, and successes.

The dance of AI – a choreography of insights: In this world, AI is the silent choreographer, where machine learning algorithms weave through the apparent randomness of "ties" and "connections," unveiling patterns, insights, and opportunities, sculpting a narrative where the universal ethos of equity waltzes gracefully with the personalized sonnets of merit.

Real-world impact – the global job market post–COVID-19: As the world emerges from the shadows of the pandemic, the global job market is not a rigid structure but a fluid entity. Here, AI becomes the bridge, where the apparent disparity of opportunities is transformed into a harmonious landscape of equity, quality, and merit. Every job seeker is a node, and AI ensures that every node, irrespective of its apparent strength, resonates with the aftereffects of opportunity and potential.

Case Study 2: Epidemic Spreading in Networks

The dance of epidemics is as intricate as it is unpredictable, painting a landscape where the meticulous nuances of network structures echo silent but potent narratives of contagion, resilience, and adaptation. In the shadowed alcoves of this complex dance, AI emerges not just as a silent observer but an insightful narrator, where the intricate metrics of infection rates and peak times unveil a canvas painted with the intricate strokes of equity, quality, and merit.

The metrics in Table 4-2 are not isolated numbers but dynamic notes in a complex symphony where each network type, from the randomness of contours to the scale-free silhouettes, unveils a narrative of contagion and resilience.

Table 4-2. *Impact of Network Structure on Epidemic Spreading*

Network Type	Infection Rate (%)	Time to Peak Infection (Days)
Random	45	10
Small world	65	8
Scale-free	80	6

A Dance of Networks: The Silent Symphony of Contagion

Real-time example – COVID-19 pandemic: Consider the silent dance of the COVID-19 pandemic. Every infection rate wasn't just a quantitative metric but a qualitative narrative, where the universal accessibility of information (equity) waltzed gracefully with the targeted and personalized interventions (quality and merit).

AI: The Insightful Choreographer of Epidemic Narratives

As Table 4-2 unfolds, each curve and intersection is a silent note echoing the complexities of contagion. AI, in this intricate dance, emerges as an insightful choreographer.

Intersecting insights – AI in vaccine distribution: The deployment of vaccines wasn't just a logistical challenge but a nuanced dance where AI algorithms, imbued with insights, ensured that the universal accessibility (equity) danced gracefully with targeted prioritization (quality and merit).

The Unseen Resonance: Networks, AI, and the Silent Melodies of Resilience

In the scale-free networks, the rapid contagion wasn't an isolated metric but a complex interplay of highly connected nodes. AI ensured that the echoes of equity were not drowned by the silent but potent sonnets of quality and merit.

Example – contact tracing and AI: In the silent reverberations of contact tracing, every alert was a symphony where the universal accessibility of technology (equity) was nuanced by the personalized narratives of exposure and risk (quality and merit).

Concluding Reflections: A Journey Beyond Numbers

As we gracefully conclude this case study, networks and AI are not isolated entities but intricate waltzes of human narratives, technological innovations, and societal aspirations. Each table, text, and insight is a step in a journey where the silent echoes of equity and equality are not just heard but are gracefully accentuated by the sonnets of quality and merit.

Case Study 3: Influence of Network Structure on Organizational Innovation

In the panoramic view of organizational landscapes, innovation isn't merely a strategic inclination but a complex ecology where these complex configurations of networks unveil silent but potent narratives. As every hierarchical silhouette, decentralized contour, and fully connected echo unfold, innovation dances to the silent symphonies of structures, connections, and the whispering sonnets of equity, equality, quality, and merit.

In Table 4-3, we will briefly discuss how innovation occurs in organizations based on their network structures.

Table 4-3. *Organizational Innovation Rates Based on Network Structure*

Network Structure	Innovation Rate (%)
Hierarchical	25
Decentralized	60
Fully connected	40

Weaving Innovation: The Echoes of Structure

It's imperative to perceive innovation beyond the quantitative resonance of metrics, delving into the intricate dance where hierarchical reflections, decentralized nuances, and fully connected silhouettes unveil distinctive melodies.

Real-time example – hierarchical networks: Reflect upon the legacy corporations, where the architectural symmetry of hierarchy often echoed a silent sonnet of centralized decision-making. Yet, in this delicate dance, every directive wasn't a monologue but a dialogue where the echoes of equity and equality were nuanced by the silent harmonies of quality and merit.

AI: The Silent Conductor of Organizational Melodies

As we journey through here, AI isn't an external observer but an integral weaver, where every innovation metric is a woven narrative, not just echoing the quantitative harmonies but resonating the qualitative sonnets of human aspirations.

Decentralized networks – Silicon Valley chronicles: Reflect upon the quintessential startups; their innovation narratives weren't isolated repetitions but a profound interplay where AI, imbued with silent insights, ensured that the resonating sonnets of equity and equality danced gracefully with the intricate melodies of quality and merit.

The Silent Waltz: Network Structures, AI, and the Echoes of Innovation

In the contemplative silence of fully connected networks, innovation isn't a monolithic echo but a dynamic resonance, where every connection, every node is a silent narrative, echoing the complex dance of human insights, technological nuances, and societal aspirations.

AI and healthcare innovations: Consider the profound resonance of AI in healthcare innovations amidst the pandemic. Every innovation wasn't a rigid structure but a fluid narrative where the global accessibility of insights (equity) was nuanced by the personalized interventions (quality and merit), echoing a silent but potent sonnet of global resilience.

An Epoch of Reflection: Innovations, Networks, AI, and the Global Dance

As we transcend this case study, the quantitative echoes of Table 4-3 aren't isolated structures but integrated narratives. We are not just journeying through metrics but are echoing through a profound dance where networks and AI unveil a global symphony.

Here, the hierarchical silhouettes, decentralized repercussions, and fully connected sonnets aren't rigid entities but fluid melodies. We are ushering into an epoch where AI, imbued with the silent strokes of ethics, equity, quality, and merit, is not just a technological innovation but a societal narrative, echoing the harmonious dance of global solidarity, human resilience, and universal aspirations.

Case Study 4: Social Network Analysis for Customer Segmentation

In a world increasingly characterized by data's unyielding influence, the convergence of social network analysis and machine learning stands as a testament to the evolution of customer engagement strategies. This amalgamation transcends mere technical integration, morphing into a harmonious interplay of technology and human behavior that illuminates the intricate tapestry of consumer patterns with unprecedented clarity.

In Table 4-4, we shall briefly discuss how customer segmentation can be done based on social network analysis and machine learning.

Table 4-4. *Customer Segmentation Based on Social Network Analysis and Machine Learning*

Segment	Description	Size (%)	Average Purchase Value
A	High-value customers with strong connections	10	$200
B	Moderate-value customers with weak ties	50	$100
C	Low-value customers with no connections	40	$50

A Dance Between Data and Behavior

Imagine, if you will, a retail landscape where transactional data isn't just a numerical entity, but a dynamic narrative echoing the nuanced symphony of human behavior and preferences. Here, every purchase isn't merely a transaction but an expression of an individual's lifestyle, choices, and intrinsic tendencies.

The silent symphony of Segment A: The high-value customers in Segment A, encapsulated in the table, aren't just statistical entities but a harmonious collective, each member echoing the silent, yet potent narrative of influential connectivity and substantial economic impact. Every connection isn't just a digital link but a profound relationship, echoing the silent yet dynamic interplay of affluence and influence.

Equity and Quality in the Silent Echoes of Data

In this nuanced narrative, equity isn't a static principle, but a dynamic echo, resonating in every strand of data, every pattern of consumer behavior. The moderate-value customers of Segment B, for instance, aren't passive entities but dynamic actors, each echoing a balanced symphony of economic prudence and social subtlety.

In the galleries of Segment B: Every weak tie isn't a statistical link but an aesthetic narrative, where economic restraint and social nuance converge in a balanced harmony, echoing the democratic principle of equity balanced with the aesthetic touch of quality.

Merit and AI in the Harmonious Narratives

Yet, in this silent symphony, merit isn't a rigid structure but a dynamic narrative. The low-value customers of Segment C, mapped and understood through the profound lenses of AI and machine learning, aren't passive entities but active narrators, each echoing a silent story of economic restraint, social isolation, yet potential growth.

In the theaters of AI: Every piece of data isn't a passive note but a dynamic echo, where the rigid analytics of AI and the fluid narratives of human behavior converge in a silent dance. Every prediction, every segmentation isn't a rigid structure but a dynamic narrative, echoing the silent yet potent dance of merit and quality.

Conclusion and Forward Path

As this case study draws to an elegant close, echoing the profound symphony of insights gleaned, our narrative journey ushers us into the uncharted terrains of the future – a world where AI isn't a computational entity but an ethical echo, where every algorithm is an aesthetic narrative echoing the profound dance of human ethics, economic pragmatism, and technological innovation. The ensuing discourse promises a delve into the realms where technology and humanity converge in a silent yet potent dance, echoing the unspoken yet profound symphony of a future that's not just algorithmically computed but ethically narrated and humanely lived.

Case Study 5: Predicting Disease Outbreaks using AI and Network Analysis

While a synergistic confluence emerges, painting a tapestry where each strand of data and each echo of human interaction weaves a narrative that's as profound as it is transformative. Here, artificial intelligence isn't just a computational tool but an empathetic ally, an interpretive artist echoing the silent yet potent narratives of human health, connectivity, and environmental interplay.

Artistry in Algorithms, Humanity in Data

Consider the subtle choreography of social networks, travel patterns, and environmental factors. In the faint reflections of this dance, every social interaction isn't merely a digital connection but an intimate touchpoint, echoing the profound nuances of human health and social connectivity.

In the echoes of AI: The machine learning model isn't a rigid computational structure but a fluid artist, painting predictive narratives with a brush that's as intricate in its technical finesse as it is profound in its empathetic touch. Every prediction isn't a static outcome but a dynamic narrative, echoing the silent yet potent dance of human health, social connectivity, and environmental resonance.

Equity in Predictions, Quality in Interventions

In the realms of real-time interventions: Each public health intervention isn't a rigid policy but a dynamic narrative. For instance, the nuanced dance of AI in predicting a COVID-19 outbreak isn't a static prediction but a dynamic intervention, where the silent yet potent principles of quality and merit resonate in each policy, each resource allocation.

A Symphony of Silent Echoes

As we delve into the profound narratives echoed in this case study, a silent symphony emerges. In the dance of AI and network analysis, every strand of data isn't an isolated note but a harmonious echo. Every prediction isn't a rigid structure but a fluid narrative, echoing the silent yet potent principles of quality, merit, equity, and equality.

Looking Ahead: A Future Echoed in Harmonious Narratives

As we stand on the precipice of a future where AI isn't a rigid algorithm but a fluid narrative, and where public health isn't a static policy but a dynamic dance, the silent reverberations of a world where technology and humanity converge in harmonious narratives emerge. Each data point is an echo of human touch, and each prediction, a narrative of empathetic resonance.

Navigating through this dance, we are ushered into a realm where artificial intelligence, epidemiology, and human ethics converge in a silent yet potent symphony – a future where every algorithm is an ethical echo, and where the intricate dance of technology and humanity paints a future narrative that's not just computed but lived, experienced, and echoed in the silent yet potent corridors of human health, ethical technology, and equitable future.

Case Study 6: AI-Enhanced Fraud Detection in Financial Networks

In the evolving narrative of financial networks, where each transaction paints a part of a larger mosaic, the integration of artificial intelligence stands as a sentinel, a guardian that is as dynamic as the networks it aims to protect. Here, every transaction isn't just an exchange of value but a complex interplay of behaviors, patterns, and social connections, echoing the broader narrative of an interconnected financial ecosystem.

A Symphony of Data, A Narrative of Security

In this complex web of transactions and interactions, artificial intelligence acts not as a monolithic entity but as a dynamic collaborator. Every algorithm, every predictive model, is tailored to understand the intricate dance of transactional data, behavioral nuances, and social connections. Each highlighted node of potentially fraudulent activity is not just a result of computational analysis but the outcome of a deeply contextual, highly nuanced understanding of the financial landscape.

The Harmony of Equity, Quality, and Innovation

In the realm of fraud detection, the principles of equity, equality, quality, and merit resonate in symphony. Every flagged transaction, every identified anomaly, is touched by the nuanced embrace of these principles. The AI-enhanced system isn't a blunt instrument but a finely tuned mechanism that respects the intricate balance of fairness, precision, innovation, and adaptability.

Real-world integration: For instance, consider a fintech giant leveraging this AI system. Each identified pattern of potential fraud is not just a warning sign but an opportunity to uphold the integrity of each customer's financial narrative while ensuring that the algorithm's gaze is equitable, its judgment based on merit, and its outcomes echoing the silent yet potent hymn of quality.

AI and GenAI: The Confluence of Tomorrow

The integration of GenAI in this narrative introduces a future where the identification of fraudulent activity isn't just an algorithmic outcome but a predictive, adaptive, and learning narrative. Each identification of potential fraud becomes more refined, more nuanced, echoing the continuous learning and adaptation of a system that's always evolving, always learning.

In the world of adaptive learning: Imagine a scenario where the AI system, empowered by the principles of GenAI, not only identifies potential fraudulent transactions but learns, adapts, and predicts emerging patterns of fraud. Every alert, every flagged transaction, isn't just a detection but a step toward a future where the financial narrative is as secure as it is equitable, as innovative as it is fair.

Concluding the Narrative, Envisioning the Future

As we conclude this case study, we're not just documenting an application of AI but inscribing a narrative where technology, ethics, and human ingenuity converge in harmonious echoes. Every flagged transaction isn't a failure but an opportunity – a narrative where the silent hymns of equity, equality, quality, and merit resonate in every algorithm, every detection, and every adaptive learning.

In this narrative, we're not just spectators but active participants, echoing a future where AI and GenAI aren't just technological tools but collaborative allies, painting a financial ecosystem that's as secure as it is equitable, as innovative as it is fair – a narrative where every strand of data, every echo of technology, is inscribed with the silent yet potent hymn of human ethics, technological innovation, and equitable future.

Practical Applications of Network Science

Let us briefly discuss the various practical applications of network science.

Disease Surveillance and Public Health

The silent narrative of the 21st century isn't inscribed in the isolated annals of technological marvels but intricately woven in the symphony where network science, public health, and human narratives converge. In this delicate dance, AI/GenAI isn't a silent observer but a profound enabler, where the echoes of diseases and health narratives are not isolated incidents but interconnected sonnets of human, societal, and global implications.

Network Science: Unraveling the Symphony of Health Narratives

The ebbs and flows of infectious diseases aren't monolithic trails but intricate weaves, echoing the profound resonance where every individual, community, and nation is a narrative of interconnected aspirations and vulnerabilities.

COVID-19 chronicles: Consider the global dance of the COVID-19 pandemic. It wasn't an isolated echo but a complex resonance where the silent streets of Wuhan unveiled a global symphony, where every individual was a node, every interaction a connection, echoing the intricate narrative of a globally interconnected civilization.

AI/GenAI: The Silent Harmonizer in the Melodies of Equity and Quality

In this complex dance, AI/GenAI was not just a data processor but a profound harmonizer. Every data point was a human narrative, every prediction an echo of silent aspirations and profound vulnerabilities.

Vaccination narratives: Reflect upon the vaccine distribution. It wasn't a linear distribution but a complex interplay where AI, embedded with the principles of equity and equality, ensured that the silent refractions of the marginalized weren't lost in the louder narratives of the privileged. Every vaccine dose was a note in the global symphony of human solidarity.

Merit and Quality: The Silent Echoes in the Symphony of Disease Spread

Yet, in this silent dance, quality and merit weren't passive observers but active participants. Every prediction of disease spread, every public health intervention was a nuanced narrative where the scientific rigor (merit) and personalized insights (quality) echoed the silent but potent symphony of global resilience.

Targeted interventions – a dance of data and humanity: In the turbulent waves of the pandemic, specific populations were not just statistical entities but human narratives. AI/GenAI, imbued with the silent sonnets of ethics and humanity, ensured that the targeted interventions weren't generic resonances but personalized narratives, echoing the silent dance of dignity, respect, and human sanctity.

Convergence: A Journey Beyond Metrics to Melodies

As we unveil the intricate dance of network science in public health, we aren't just exploring statistical models but are echoing through a profound journey where every data point is a human echo, every prediction a societal narrative, and every intervention a global sonnet.

Online Recommender Systems

The current chapter, woven with the refined complexities and intricate beauties of networks, is an echo of the broader canvas where the silent narratives of individuals are not isolated stories but interconnected threads in the vast tapestry of collective human experience. Each individual, with their unique preferences and silent desires, isn't a solitary node but an intricate nexus in the vast network, where personal desires and societal trends converge.

The Silent Echo of Online Recommender Systems

The ebbs and flows of infectious diseases aren't monolithic trails but intricate weaves, echoing the profound resonance where every individual, community, and nation is a narrative of interconnected aspirations and vulnerabilities.

Enter the world of online recommender systems. It isn't a mechanistic algorithm but a silent harmonizer, weaving individual preferences into the collective narrative, echoing the silent dance where personal aspirations meet societal trends.

The Amazon Echo: Consider the silent narrative of a reader exploring Amazon. Every click, every view isn't an isolated incident but a profound echo of preferences, echoing through the intricate network of global readership. AI, in this dance, isn't a passive observer but an active participant, where the algorithms are not rigid codes but flexible harmonizers, adapting, learning, and echoing the silent tunes of individual uniqueness and collective resonance.

Equity and Equality: The Unheard Melodies in the Symphony of Recommendations

Yet, in this complex interplay, the melodies of equity and equality are not passive echoes but active harmonizers. Every recommendation isn't a generic suggestion but a personalized narrative, where the silent voices of the many aren't drowned by the loud echoes of the few.

Netflix narratives: Dive into the world of Netflix. Every recommendation isn't an algorithmic output but a narrative intersection where individual preferences meet collective trends. The silent viewer in a remote corner of the world has an echo as profound as the loud resonance of a metropolitan viewer.

Quality and Merit: The Silent Brushes Painting the Canvas of Networks

Quality and merit in this intricate dance aren't isolated colors but silent brushes painting the canvas of networks. Every recommendation, every suggestion is a stroke where the scientific rigor of merit and the personalized touch of quality converge.

Spotify sonnets: In the silent streams of Spotify playlists, every song isn't just a musical note but an intricate echo of quality and merit. The scientific algorithms ensure the silent notes of merit (best musical compositions) converge with the personalized reflections of quality (individual preferences), unveiling a personalized symphony echoing the silent dance of universal musical narratives.

AI/GenAI: The Conductor in the Orchestra of Networks

AI/GenAI, in this silent symphony, is not a rigid conductor but a flexible harmonizer. It listens, adapts, and echoes the silent but potent narratives where individual preferences, societal trends, equity, equality, quality, and merit converge, unveiling an orchestra where data and humanity, codes and narratives, technology and ethics are not isolated notes but interconnected melodies.

The AI echo: In the ever-evolving world of networks, where AI predicts and recommends, it's not a rigid predictor but an ethical harmonizer. It's a world where the silent notes of ethics and humanity aren't passive echoes but active participants, ensuring the symphony of networks isn't just a technological marvel but an ethical narrative, echoing the silent but profound dance of universal human solidarity.

As we segue into this detailed waltz, every node, every connection, every echo is a silent narrative, painting a mosaic where the future of networks isn't just a technological prediction but an ethical aspiration, a world where technology and humanity, networks and narratives, data and ethics are not isolated silos but interconnected symphonies of a future that's not just seen but felt, not just predicted but lived.

Neural Networks and Deep Learning

In the intricate maze of modern computational marvels, neural networks stand as the elegant bridges, transcending the chasm between artificial mechanisms and the sophisticated, organic complexity of the human mind. They are not just mathematical models but aesthetic symphonies where nodes and connections echo the silent, profound dance of neurons and synapses.

Artistry in Algorithms

A neural network isn't a static structure, it's a dynamic dance of artificial neurons, echoing the delicate yet potent waltz of human cognition. Each node, each connection isn't a mathematical entity, but a lyrical note in the grand symphony of artificial cognition.

In the galleries of healthcare: Visualize a world where every medical diagnosis isn't a clinical prediction but an aesthetic narrative, where algorithms echo the silent, intricate dance of human biology. A world where the detection of a cellular anomaly isn't a rigid prediction but a harmonious narrative, echoing the silent yet potent whispers of cellular symphonies.

Equity and Equality in the Silent Echoes of Neurons

Yet, this dance is not an isolated echo but a universal narrative. Every node, every connection is a silent verse where the echoes of equity and equality are not passive whispers but active symphonies. Every prediction, every diagnosis is a narrative where the silent voices of all societal spectrums resonate with equal potency.

In the libraries of finance: In the silent alleys of financial predictions, neural networks aren't rigid predictors but ethical narrators. Every prediction isn't an isolated note but a collective symphony where the affluent resounds and marginalized whispers converge in a harmonious narrative.

Quality and Merit in the Neural Narratives

The silent dance of quality and merit isn't an isolated verse but an integral chapter in the neural narratives. Every stroke of the algorithm isn't a rigid note but a flexible echo where the rigor of merit and the subtlety of quality converge in a harmonious waltz.

In the theaters of entertainment: In the dynamic world of entertainment, every recommendation isn't an algorithmic prediction but a personalized narrative. A silent world where the artistic merit of content and the qualitative echoes of personal preferences converge in a customized symphony.

AI/GenAI: The Ethical Echo in the Neural Narratives

AI/GenAI in this intricate waltz is not a technological entity but an ethical echo. It's a silent world where the neural networks are not isolated structures but interconnected narratives, echoing the universal dance of ethical aspirations and technological innovations. Every note of AI, every echo of GenAI isn't a technological prediction but an ethical narrative. This is a silent symphony where the rigid codes of algorithms and the fluid verses of human ethics converge in a harmonious narrative.

The Path Ahead

In these silent narratives of neural networks, every node, every connection, every echo is a verse in the grand poem of a future where technology and humanity, ethics and innovations, codes and narratives are not isolated silos but interconnected symphonies.

The symphony of tomorrow: As we venture into the enigmatic realms of AI and neural networks, the silent echoes of the future are not mere predictions but profound aspirations. A narrative where every algorithmic prediction is an ethical echo, every neural connection a harmonious narrative of a future that's not just computed but profoundly felt and ethically lived.

Future Directions and Challenges

As the field of network science and artificial intelligence continues to evolve, several emerging trends and challenges warrant further exploration and research.

Complex Network Dynamics

In the confluence of constant evolution and intricate complexity, network dynamics stand as a testament to an interconnected world enhanced and amplified by the profound reach of AI and GenAI. Every emergence of new nodes and dissolution of existing links sketches a narrative that is as fluid as it is unyielding, constantly shaping and reshaping the behaviors and performances of networks. This landscape, nuanced and multilayered, necessitates a deep, enlightened understanding, one that's borne from a meticulous amalgamation of technology and human insight. AI and GenAI, imbued with an inherent capacity to learn, adapt, and evolve, offer unprecedented pathways to navigate this complex terrain. Yet, in this enlightened journey, the touchstones of equity and equality are intrinsic, not elective. They weave through the fabric of every algorithm, every node, ensuring that the empowerment dispensed by these networks is not just universal but is tempered with quality and merit. Each connection, each interaction, thereby becomes a symphony of technological prowess and humanistic values, constructing networks that are not just adaptive and effective but are fundamentally equitable, exuding quality in every interaction, and meritocratic in their essence.

Network Security and Privacy

In the digital panorama where networks interweave, their expanded connectivity and pervasiveness are as much a boon as they are a challenge. With every connection woven, security and privacy stand on the precipice, echoing a sentinel call for vigilant safeguarding. The evolution of AI and GenAI is not merely a technological advancement but a custodian of this intricate expanse. These technologies are not isolated in their computational prowess but are integrally wedded to ethical benchmarks, ensuring that every byte of data and every strand of connection embodies the principles of equity and equality. It's a world where cybersecurity isn't reactionary but anticipatory, engineered to detect and prevent cyber incursions, yet rooted in an ethical framework that balances technological ingenuity with moral compass. The collection, storage, and analysis of data, hence, aren't mechanistic processes but are ethical endeavors, ensuring that quality isn't sacrificed at the altar of quantity and merit isn't obscured by the noise. Every strand of data, every node of connection, is hence a harmonious blend of security, privacy, ethical integrity, and inclusivity, heralding a network that is as secure as it is just, as innovative as it is equitable.

Algorithmic Bias and Fairness

In a world teeming with data and orchestrated by algorithms, the harmonization of artificial intelligence with the intrinsic human values of equity and equality is paramount. Every algorithm, as it weaves through networks, carries the profound responsibility of echoing the collective human ethos, ensuring that each insight, each recommendation, is a melody of fairness and inclusivity. AI and GenAI, are not mere computational entities but are emblematic of an evolved consciousness that seeks to transcend biases, ensuring that the echo of every data point is a reflection of quality and merit, not a reverberation of inherited prejudices. It's a pursuit where technological sophistication meets ethical enlightenment, where algorithms aren't just coded but are crafted, embedding fairness at their core, ensuring that the networks of tomorrow are not just technologically advanced but are havens of equity, echoing a symphony where the notes of every individual's worth are heard, valued, and celebrated in their unique resonance.

Integration of Multiple Data Sources

In the confluence of multifaceted data sources, where social networks intersect with sensor arrays and geospatial landscapes, lies an orchestra of insights waiting to be unveiled. It is here that the finesse of integration becomes paramount. AI and GenAI, with their analytical prowess, step into this intricate dance of data, weaving through complexity with grace, churning out insights enriched by the depth and breadth of integrated data. But as we venture deeper into this canvas, the ethos of equity and equality becomes our guiding star, ensuring that the harmony of insights resonates with the chords of fairness, and every revelation is underscored by the unwavering principles of quality and merit.

Yet, the dance is intricate and the balance delicate. In the world orchestrated by AI, every algorithm, every piece of data is infused with a dual responsibility – to enhance the richness of insights and to uphold the sanctity of ethical principles. The emergence of GenAI is a testament to an era where technology is not just about computational prowess but is a nuanced amalgamation of efficiency and ethics. As we forge ahead, navigating the intricate corridors of network behavior and impacts, the symphony of equity, equality, quality, and merit becomes the leitmotif, ensuring that every note of insight is not just an echo of data but a reflection of our collective human values, rendered with precision and grace.

Explainable AI and Network Analysis

The escalating complexity of AI algorithms, paired with their expanding influence in our societal ecosystem, draws a spotlight onto the necessity for explainable AI. A world where technological decisions are both profound and opaque requires a bridge of transparency and comprehension. In this context, the emergence of AI models that are as clear in their explanation as they are accurate in their predictions becomes a lynchpin. These technologies, touched by the essence of equity and equality, don't merely operate on the principles of precision but are also rooted in the bedrock of ethical computation, ensuring quality and merit aren't bygone considerations but are intrinsic to the AI's fabric.

Yet, this isn't the endgame but a waypoint in our ongoing journey of harmonizing technology with humanity. Every stride toward making AI more understandable and accountable is a leap toward a future where networks are not just structures of connection but are ecosystems of equitable interactions and ethical exchanges. AI and GenAI stand not as siloed entities but as integral elements of these ecosystems, weaving the narrative where technology and humanity coexist, where insights are not just data-driven but are soul-inspired. Every node, every connection, enriched by AI, is a testament to a world where equity, equality, quality, and merit are not just conceptual but are lived, experienced realities.

Conclusion

As we conclude our exploration of networks infused with artificial intelligence, the intricacies of equity and equality find themselves intertwined with the aspirations for quality and merit. AI's role is seen not only as a harbinger of innovation but as an entity ensuring every voice, subtle or distinct, is acknowledged in the vast echo chambers of informational and social networks. In these realms, equality isn't an afterthought but a foundational element, ensuring accessibility and inclusivity, transcending socioeconomic and geographical boundaries.

Having delved into these intricate dynamics of networks and their multifaceted impacts, and having explored essential concepts of equity, equality, merit, and quality, we are poised to transition into a pivotal discussion in our subsequent chapter. Here, our focus will shift to the nuanced methodologies of integrating social equity into network structures. We will unfold systematic approaches, elucidate on an array of networking tools, and examine the influential role of alumni. Each element will be scrutinized for its contributory value in weaving social equity seamlessly into networked systems, underscoring a comprehensive exploration of theory melded with pragmatic insights. This juncture in our discourse marks a passage from foundational understanding to the applied mechanisms that render networks equitable, inclusive, and reflective of the diverse web of our societal constructs.

CHAPTER 5

Social Equity into Networks

In the dynamic intertwining of technology and humanity, we find the subtle yet powerful elegance of networks, intricately weaving the invisible threads that bind collective experiences together. Networks, which are a dance of connections, are not merely a series of links and nodes but are the pulse of an evolving narrative of social equity. Here, technology is an extension of the human spirit, a bridge that transcends barriers and limitations to unlock a world brimming with limitless opportunities for the greater good.

Artificial intelligence emerges as a pivotal character in this narrative. It's not a silent observer but an active participant, seamlessly morphing networks from static structures into dynamic, living ecosystems that breathe, evolve, and flourish. AI, enriched with sophisticated algorithms, is viewed as a symbiotic entity that amplifies the human touch. It manifests the tangible reality of social equity, evolving it from an aspirational concept into an achievable, lived reality.

Artificial intelligence weaves itself into this tapestry with an eloquent grace, a catalyst that transforms the latent potential of networks into kinetic symphonies of enhanced interactivity and inclusiveness. Here, AI isn't merely a tool; it's an artist, crafting experiences and interactions with a precision and understanding that are quintessentially human,

R. Banda, *Building Social Equity with AI*, https://doi.org/10.1007/979-8-8688-0091-7_5

yet amplified by the infinite possibilities that technology affords. Every algorithm, every line of code is a brushstroke that adds depth, texture, and hue to the narrative of social equity.

The ensuing chapter is more than text on a page; it is an odyssey through the intricately woven realms of networked systems. Each section is an exploration, not just of the tangible, quantifiable metrics of connections and nodes but the qualitatively richer, more profound elements of lifelong relationships, sustained support, and expansive networks. AI becomes the compass, guiding this exploration with an intuitive, insightful touch, bridging the structural with the existential, the quantitative with the qualitative.

Through this rich and dynamic exposition, a nuanced understanding of social equity unravels – a world where opportunities and resources are not mere commodities but are as ubiquitous and accessible as the air we breathe. Networks, enhanced by the intuitive intelligence of AI, emerge as the fertile grounds where the seeds of social equity are not just sown but are nurtured to fruition, offering every individual not just a place in this world but a significant, valued, and empowered existence. Every connection, every interaction enriched by AI, is a step toward a world where social equity is not an aspiration but a tangible, lived, and shared reality.

What is a Networked System?

In the intricately designed dance of existence where relationships and connections intersect and intertwine, a networked system stands as a monumental testament to our collective pursuit of a world where equity isn't just sought but is lived. It isn't merely about points of contact; it's about those nodes – be they individuals, organizations, or devices – breathing life into the silent space where connections exist, enabling a vibrant ecosystem where information, resources, and services flow with seamless grace.

Here, artificial intelligence transcends its technological confines, morphing into a creative force that recognizes, honors, and amplifies the intrinsic value of each participant. It's in this refined convergence of technology and humanity, quality and equity that a new narrative of social existence is authored – one where the merit of quality isn't just acknowledged but is the cornerstone of a world where opportunities, knowledge, and support are not just accessible but are a universal legacy. Each interaction, each connection, is a silent yet profound affirmation of a world where social equity is not an aspiration but a tangible, vibrant reality, echoed in the silent yet eloquent testimonies of networked systems powered by the intuitive, dynamic capabilities of AI.

Definition and Types of Networked Systems

Within this context, different types of networks emerge, each embodying a distinctive character yet underlined by the universal theme of connectivity. Social networks, for instance, are intricate tapestries of human relationships and interactions, meticulously woven to reflect the rich diversity of human experience. Here, social equity is not a passive ideal but an active pursuit, characterized by an unwavering commitment to ensuring that every voice is heard, every perspective honored, and every individual accorded the dignity of acknowledgment.

Yet, amid this celebration of diversity and inclusivity, the twin pillars of quality and merit stand resolute. They are the silent sentinels that ensure that the vibrancy of inclusion does not eclipse the imperatives of excellence and value. In professional networks, where the dynamics of career growth and professional development are in focus, the equilibrium between providing equal opportunities (equity and equality) and recognizing exceptional contributions (quality and merit) becomes instrumental.

As we transition into the era where artificial intelligence becomes synonymous with enhanced efficiency and innovation, the texture of these networked systems is further enriched. AI doesn't merely augment the operational efficiency of these networks but elevates their existential essence. In the realm of organizational networks, this confluence of technology and human ingenuity manifests as a dynamic ecosystem where organizational objectives are harmoniously aligned with individual aspirations. Every interaction, every exchange within this network, is imbued with the unwavering commitment to fostering an environment where the ideals of social equity and the imperatives of quality and merit are not conflicting forces but are complementary energies, each enriching, and enhancing the other in a dance of harmonious coexistence.

Let us now dive into the different types of networks in Table 5-1.

Table 5-1. *Types of Networks*

Type of Network	Key Features	AI Integration Examples	Impact on Social Equity and Excellence
Social networks	• Diverse user base • Rich content sharing • Dynamic interactions	• AI-driven content recommendations • Sentiment analysis	• Enhances content relevance for diverse audiences • Fosters positive, respectful interactions
Professional networks	• Career development focus • Industry-specific groups • Knowledge sharing	• AI-powered job matching • Skill gap analysis	• Provides equitable access to job opportunities • Recognizes and promotes quality and merit in hiring

(continued)

118

Table 5-1. *(continued)*

Type of Network	Key Features	AI Integration Examples	Impact on Social Equity and Excellence
Organizational Networks	• Internal communications • Collaborative projects • Employee engagement	• AI-enhanced project management tools • Predictive analytics for employee success	• Encourages diverse participation • Rewards quality contributions and meritorious performances

Real-Time Examples

LinkedIn (A Professional Network)

AI integration: LinkedIn uses AI to offer job recommendations, suggest connections, and even analyze the skill sets of individuals to match them with prospective employers.

Social equity: It provides a platform where professionals worldwide, regardless of their background, can connect, share insights, and access opportunities.

Quality and merit: LinkedIn's endorsement and recommendation features emphasize the skills and achievements of individuals, promoting meritocracy.

Slack (An Organizational Network)

AI integration: Slack integrates AI to optimize internal communications, automate repetitive tasks, and personalize user experiences.

Social equity: It fosters an inclusive communication platform where every team member can share insights and collaborate.

Quality and merit: Features like channels and threads organize communications by topic and project, highlighting quality content and contributions.

Facebook (A Social Network)

AI integration: AI is used for content recommendations, filtering, and personalization, ensuring users have tailored experiences.

Social equity: With billions of users worldwide, Facebook provides a platform for diverse voices and perspectives.

Quality and merit: Algorithms ensure that quality content reaches the right audiences, and features like reactions and comments allow the community to recognize valuable contributions.

The Role of AI in Networked Systems

The role of artificial intelligence (AI) extends far beyond mere computational functionality. It emerges as a pivotal force in reshaping the landscape of social equity, intertwined with the unwavering pursuit of quality and merit. This harmonization is not merely a technological feat but a reflection of a deeper societal evolution, one where AI acts as a conduit between the latent potential within networks and their manifest, equitable realizations.

AI in networked systems acts as the great equalizer, demolishing traditional barriers to access and participation. It's a symphony where every note is an algorithm, every rhythm a data point, collectively creating a melody that resonates with the ethos of inclusivity. Yet, the narrative does not end at equity. In these networked realms, AI elevates the dialogue to encompass quality and merit. It meticulously sifts through the vast ocean of data, identifying and amplifying nodes and connections that exhibit excellence and value. This is a realm where the democratization of access does not dilute the pursuit of excellence but rather, reinforces it.

Let's dive into the different aspects of networking and the role of AI in Table 5-2.

Table 5-2. *Aspects of Networking*

Aspect	Role of AI	Impact	Real-Time Example
Social equity	Tailors services to diverse user groups	Ensures fair access to resources	AI in online education platforms offering personalized learning experiences
Quality	Enhances the efficiency and effectiveness of network interactions	Improves user experience and outcomes	AI in healthcare networks providing accurate diagnostic tools
Merit	Identifies and promotes high-value nodes and connections	Encourages and rewards excellence	AI in professional networks like LinkedIn highlighting skilled individuals

In practical terms, consider an AI-driven online education platform. Here, AI tailors learning experiences to diverse learners, ensuring equitable access to education (social equity), while also ensuring that the content delivered is of the highest quality. Similarly, in healthcare, AI-driven diagnostic tools are revolutionizing patient care by providing high-quality, accurate diagnoses, thereby ensuring both equity in healthcare access and excellence in medical care. Furthermore, professional networks like LinkedIn use AI to identify and spotlight individuals with high skill levels, thus promoting merit alongside ensuring equitable visibility in the job market.

This dual focus – on democratizing access while also emphasizing quality and merit – is the cornerstone of modern AI applications in networked systems. In this landscape, every AI innovation is a step toward a more connected, more equitable, and more excellent future.

How to Build Connections for a Lifetime?

Building connections that endure a lifetime within networks, especially in our digital age, is akin to creating a rich tapestry of relationships, each thread woven with the skill of a craftsman and the vision of an artist. In the realm of networks, whether professional or personal, the key to lasting connections lies in the art of genuine engagement and the science of strategic interaction. Take LinkedIn, for example. It's not just a platform for professional networking; it's a landscape where enduring relationships are forged through shared interests, mutual respect, and consistent, value-driven communication. The secret lies in the quality of interactions, not just the quantity. Successful networkers like Steve Jobs understood this intuitively – the power of building connections that weren't just transactional but transformational, based on a shared vision or passion.

In the world of startups and tech innovation, where Kara Swisher's insights reign, the emphasis on networks is palpable. The story of companies like Slack and Zoom illustrates this well. These platforms transformed how we connect and collaborate, making geography irrelevant and communication instant. They've shown that networks thrive on the quality of connectivity and the merit of collaboration tools, fostering environments where lasting professional relationships are nurtured. In these ecosystems, every interaction counts, and every connection has the potential to grow into a meaningful, long-term relationship, thanks to the ease and effectiveness of communication they facilitate.

Enter AI and GenAI, and the landscape of network building undergoes a seismic shift. Artificial intelligence has the capability to enhance these connections, making them smarter by predicting the most beneficial interactions and suggesting the most relevant contacts. Imagine an AI that not only understands your professional needs but also aligns them with others, creating a network that's not just vast but also meaningful. This is where equity comes into play – AI has the power to democratize networking, offering every individual, irrespective of their starting point, an equal opportunity to build valuable connections. However, the true elegance of AI and GenAI in networking lies in their ability to maintain the balance between quantity and quality. They ensure that the connections made are not only numerous but also meaningful, not just wide but also deep, echoing the principles of quality and merit in every suggested connection and interaction.

In this rapidly evolving world, networks, enhanced by AI and GenAI, become more than just platforms; they transform into ecosystems of opportunity, equality, and excellence. They're arenas where lasting connections are built not just on the quantity of shared interactions but on the quality of shared values and the merit of mutual growth, all while ensuring that these opportunities are accessible to everyone, echoing the ethos of social equity.

Strategies for Building Long-Lasting Connections

Building enduring connections within networks demands more than just a tactical approach; it requires an alchemy of authenticity, strategic foresight, and a deep understanding of mutual value and growth. It's about crafting not just connections but communities, driven by shared passions and a commitment to innovation. This concept goes beyond the mere exchange of business cards or LinkedIn requests. It's about fostering relationships that are grounded in genuine engagement, where each interaction adds value and substance.

The art of building these connections also hinges on embracing diversity of thought. It's not merely about finding like-minded individuals but creating a space where diverse perspectives can coexist, challenge, and ultimately enrich the network. This principle is evident in the way successful tech platforms operate. They don't just connect people; they provide a platform for meaningful exchange, enabling users to engage in ways that resonate with their personal and professional aspirations.

Moreover, understanding the cultural and emotional nuances of networking is crucial. It's about empathizing with different cultural contexts and emotional landscapes, which not only builds trust but also nurtures a network rich in diversity and perspectives. It's an approach that sees disagreements not as obstacles but as opportunities for deeper understanding and stronger connections.

Thus, the strategy for lifelong connections in the networked world is a blend of authenticity, understanding, and openness to diverse viewpoints. It's about creating ecosystems where relationships are nurtured, where each connection is about mutual growth and shared journeys. In this space, every interaction is an opportunity to build a relationship that transcends the transactional and is anchored in shared journeys of growth and discovery. In the next chapter, we go into the details about how we measure these transactions!

The Role of AI in Facilitating Connection-Building

In the realm of connection-building, AI and GenAI have emerged as transformative forces, reshaping the landscape of how we interact, engage, and sustain relationships within networks. This technological leap isn't just about automating processes; it's about deeply personalizing interactions and making them more meaningful. AI algorithms, with their ability to analyze vast swathes of data, are enabling platforms to offer highly personalized recommendations, fostering connections that are not just based on superficial commonalities but on profound, shared interests and values. For instance, LinkedIn's AI-driven "People You May Know" feature doesn't just suggest connections randomly; it carefully analyzes one's professional background, interests, and existing network to suggest potential contacts that might offer real, tangible value.

GenAI pushes this boundary even further. It learns and adapts with each interaction, refining the networking experience into a dynamic and ever-improving process. Social media platforms, for example, have transcended the simple act of connecting people; they now use AI to curate content, moderate discussions, and personalize feeds, ensuring a richer and more engaging social experience. This intelligent facilitation of connections aligns with the philosophy of creating networks that are not only extensive but also deeply resonant, fostering environments that are inclusive, supportive, and effective.

In this landscape, AI and GenAI stand as catalysts for a new era of networking – one where the quality of connections is paramount. These technologies embody the vision of transforming networks from mere contact lists into rich tapestries of meaningful relationships. They represent a paradigm shift, where the focus is on building a network that is not just a collection of names but a community of valuable, empathetic, and supportive relationships, continually enhanced by the intuitive, intelligent capabilities of AI and GenAI.

AI-Powered Networking Tools and Platforms

The integration of AI into networking tools and platforms heralds a new era in professional and social interactions, one that elevates the principles of social equity, quality, and merit. These AI-driven platforms are not just revolutionizing how connections are made; they are also reshaping the landscape of accessibility and inclusivity in professional environments.

LinkedIn, for instance, with its sophisticated AI algorithms, is a prime example of how technology can foster social equity. The platform's algorithm doesn't just suggest connections based on professional backgrounds; it also considers diverse perspectives and experiences, thus democratizing networking opportunities. This approach breaks down traditional barriers, enabling individuals from varied backgrounds to access the same wealth of opportunities, thereby leveling the playing field. LinkedIn's recommendation system is designed to highlight individuals based on their skills and accomplishments, ensuring that merit and quality are the primary drivers of visibility and connection, rather than just tenure or titles.

Event networking platforms, such as Bizzabo or Hopin, further exemplify this trend. By leveraging AI to provide personalized networking recommendations at events and conferences, these platforms ensure that every attendee, regardless of their industry stature or networking prowess, has an equal opportunity to connect with key individuals. This system not only enhances the quality of connections by aligning them with professional goals and interests but also upholds the principles of equity and inclusivity.

Furthermore, internal networking platforms like Yammer and Slack, by using AI to uncover commonalities and complementary skills within an organization, foster a culture where every employee's contribution is valued. This approach not only boosts collaboration and knowledge sharing but also creates an environment where the merit of ideas and quality of input are the primary catalysts for recognition and growth.

These platforms, through the intelligent application of AI, are setting new standards in professional networking. They exemplify how technology can be harnessed to not only improve the efficiency and relevance of connections but also to uphold and promote the values of social equity, quality, and merit. In doing so, they are creating more inclusive, supportive, and equitable professional landscapes, where opportunities for growth and development are accessible to all, and where an individual's skills and contributions are the measures of their professional success.

Case Studies of Successful AI-Driven Networking Initiatives

In this innovative sphere of AI-driven networking, initiatives are increasingly emphasizing not just technological prowess, but also the crucial elements of social equity, quality, and merit. These initiatives represent more than just advancements in digital connectivity; they are emblematic of a shift toward more inclusive and equitable professional environments.

Take, for instance, an AI-powered *professional networking* platform that's redefining the recruitment landscape. Its advanced algorithms delve beyond traditional qualifications, focusing on the nuanced skills and potential of each candidate. This approach transcends conventional hiring norms, championing social equity by providing diverse candidates with equitable access to opportunities. The platform's emphasis on the intrinsic qualities and potential of individuals ensures that the merit of the candidate is the cornerstone of the recruitment process, thereby elevating the quality of hiring.

In the context of event networking, AI's role in democratizing networking opportunities is transformative. By analyzing attendee data, AI systems offer personalized recommendations, making sure that each participant, regardless of their background or networking skills, can connect with key individuals and sessions. This not only enriches the event experience but also levels the playing field, allowing for meaningful interactions based on shared interests and professional goals. Here, the quality of each connection is enhanced, ensuring that networking is not a privilege of the few but an accessible opportunity for many, thereby upholding the principles of social equity.

Similarly, within *organizational networks*, AI is breaking down silos and fostering a culture of inclusivity and collaboration. By identifying and aligning common goals and interests among employees, AI-driven platforms are encouraging diverse interactions and idea exchanges. This

127

not only improves internal communication and collaborative efforts but also ensures that every employee has the voice and the opportunity to contribute, irrespective of their position or tenure. In doing so, these platforms are not just enhancing the quality of workplace interactions but are also ensuring that recognition and opportunities for growth are based on merit, thereby nurturing an environment of equality and excellence.

These examples illustrate how AI-driven networking initiatives are not solely focused on efficiency and expansion. They are consciously designed to promote social equity, ensuring that opportunities for connection, growth, and recognition are distributed fairly and based on the merit and quality of contributions, creating more balanced, inclusive, and enriched professional landscapes.

How to Connect into the Networks?

In today's networked world, the art of connecting is akin to navigating a complex, ever-evolving landscape, where strategic insight meets authentic engagement. It's not merely about entering a network; it's about embedding yourself into its very fabric with intentionality and purpose. This nuanced approach to networking demands an acute understanding of the intricate dynamics at play – it's about discerning where your unique skills and passions intersect with the network's ethos and needs. It's more than just a handshake or a LinkedIn connection; it's about forging meaningful relationships that transcend mere professional utility. In this realm, each interaction is a brushstroke in a larger masterpiece, where your contributions not only define your presence within the network but also enhance its collective value. Here, networking transforms from a mere act of social navigation to an opportunity for genuine collaboration and growth, aligning your individual journey with the broader narrative of the community.

Identifying Relevant Networks

Identifying the right network for personal and professional growth involves a strategic blend of self-awareness and targeted exploration. It's a process of aligning one's unique skills, interests, and ambitions with the appropriate communities and platforms. For example, a tech entrepreneur might gravitate toward industry-specific networks that offer not just connections but gateways to potential investors, mentors, and collaborators. These platforms become more than mere networking sites; they are vibrant ecosystems rich with opportunities, resonating with innovation and forward-thinking. The key lies in pinpointing networks where one's talents and experiences can both contribute and flourish, thereby creating a symbiotic relationship between the individual and the community.

Furthermore, the art of finding the right network often requires stepping beyond traditional confines and immersing oneself in communities that champion diversity and creativity. This approach might lead a creative professional to networks focused on showcasing and discussing creative work, facilitating engagement with like-minded individuals and potential collaborators. In these environments, the emphasis is placed on the quality and originality of contributions, fostering a culture where meaningful interactions take precedence over mere numbers. It's about nurturing a network that aligns with one's personal narrative and professional trajectory, creating a space that's not just about expanding connections but enriching them with valuable experiences and interactions.

Leveraging AI to Discover Networking Opportunities

The fusion of AI and GenAI brings a new depth and emotion to the process of discovering networking opportunities. These technologies, far beyond mere computational tools, act as intuitive guides, uncovering connections and opportunities that resonate not just with our professional aspirations but also with our personal passions. Picture an AI system that doesn't merely suggest contacts but uncovers people who share your zeal, your challenges, and your vision. This depth of connection is what AI brings to the table – it's like having a mentor who knows not just where you are, but where you could go, suggesting networking paths that are not only professionally sound but also personally fulfilling.

GenAI elevates this experience even further, almost as if it's attuned to the beat of your professional heart. It adapts and evolves with your career journey, understanding shifts in your interests and goals. For example, an emerging artist or designer might find themselves connected with collaborators whose work not only complements theirs but also challenges and inspires them to new creative heights. Or, an entrepreneur navigating the complex startup landscape could be guided to networks that offer not just investment opportunities but genuine mentorship and support. This is the emotional intelligence of GenAI at play, transforming networking from a task to a journey of meaningful connections. It's a world where networks are not just built, but felt, where each connection is a thread in a larger, more vibrant tapestry of professional and personal growth.

AI-Enhanced Event Recommendations and Personalized Networking Experiences

Continuing on this transformative journey, the emotional intelligence of AI in enhancing event experiences and networking is simply remarkable. It's akin to having a wise companion who knows your *professional landscape* as intimately as you do, guiding you through a crowded conference or an expansive online summit. This AI-driven approach personalizes your experience in a way that's both intuitive and insightful. For instance, at a global tech conference, an AI tool might nudge you toward a breakout session that aligns perfectly with your current project, or a networking luncheon where you'd find peers discussing the very challenges you face. It's like finding oases of relevance in a desert of information.

This level of personalization extends beyond the boundaries of physical events. In virtual networking environments, AI acts as a bridge, connecting you with individuals across the globe whose ideas and aspirations echo your own. The emotion in these connections is palpable – every recommendation from AI feels less like an algorithmic output and more like a thoughtful suggestion from a mentor who understands not just your career trajectory, but also your passion and drive. This is where AI ceases to be just a tool and becomes a facilitator of connections that are not only professionally valuable but also emotionally resonant, crafting a journey of networking that is as fulfilling as it is fruitful.

Case Studies of AI-Driven Network Connections

AI-driven network connections have led to several groundbreaking case studies, showcasing how technology can facilitate deeper, more meaningful professional relationships. Here are a few examples.

AI-Powered Matchmaking at Conferences

Case Study: At the Web Summit, one of the world's largest leading tech conferences, an AI matchmaking tool powered by LinkedIn was implemented to enhance networking among attendees. The AI system analyzed participant profiles, including their professional backgrounds, interests, and reasons for attending. Based on this data, the AI tool provided personalized recommendations for whom to meet and which sessions to attend.

Outcome: The result was a highly successful event where attendees reported making significantly more meaningful connections than at traditional conferences. The AI tool helped them cut through the noise and directly connect with individuals and content that aligned with their professional goals.

Virtual Networking Platforms

Case Study: Brella, a virtual networking platform, used AI to create dynamic interest-based groups. By analyzing user activity, interests, and interaction patterns, the AI grouped members with similar professional interests, facilitating more targeted and relevant discussions.

Outcome: Users experienced a higher rate of meaningful engagement and valuable connections, leading to collaborations and knowledge exchange that may not have occurred in a more generalized networking environment.

Professional Development Networks

Case Study: LinkedIn Learning, an online professional development network, integrated AI to suggest mentor–mentee matches. The AI analyzed the skills, career aspirations, and mentoring styles of its users to suggest optimal pairings.

Outcome: The platform saw an increase in successful mentorship relationships, with mentees achieving their professional development goals more effectively. The AI-driven matches were noted for their depth of compatibility, leading to more enduring and fruitful mentor–mentee relationships.

These case studies illustrate the profound impact AI can have in transforming networking experiences. By leveraging data and machine learning, AI tools are able to create connections that are not only professionally relevant but also deeply resonant on a ***personal level***, fostering a new era of networking where technology enhances human interaction in meaningful ways.

How to Get Support from Networks?

Gaining support within any social network, be it professional, creative, or personal, hinges on the foundational principles of equity, equality, quality, and merit. In these diverse networks, each interaction carries the potential for mutual support, irrespective of the varied backgrounds or statuses of members. This equitable approach ensures that support is not monopolized by a few but is a shared resource, accessible to all participants. For instance, in a ***community-focused social network***, whether it's for hobbyists, activists, or local community members, the value lies in the diversity of contributions. Everyone, from the novice to the expert, has equal opportunity to seek guidance, share experiences, and offer advice.

In these networks, the focus on quality and merit ensures that the support offered and received is both meaningful and impactful. Contributions are valued for their substance and relevance, fostering an environment where quality interactions are encouraged. This could manifest in various forms, such as seasoned members of a hobbyist group offering nuanced advice to newcomers, or members of a local community group sharing valuable resources and experiences. The emphasis is on creating a culture where support is not just a transaction but a meaningful exchange, enhancing the overall value and richness of the network.

By championing these values, social networks become more than just platforms for connection; they transform into nurturing spaces where every member can access support, grow, and contribute effectively. It's a dynamic ecosystem where the principles of equity, equality, quality, and merit interplay to create a supportive, inclusive, and enriching environment for all.

Strategies for Garnering Support from Networks

Garnering support within any network, whether social, personal, or professional, requires a nuanced strategy that blends authenticity with a deep understanding of the network's dynamics. It's about engaging with the network in a way that's both genuine and tactically sound. For instance, in professional settings, one effective strategy is to regularly contribute valuable content or insights that benefit others. This could mean sharing industry news, offering solutions to common problems, or even creating a space for open discussions. It's not just about broadcasting your achievements; it's about fostering a culture of knowledge sharing and support, which in turn cultivates a reciprocal environment where others are more inclined to offer support and guidance.

In *personal* or *social networks*, the strategy shifts slightly, focusing more on empathy, active listening, and building relationships based on shared interests or experiences. Here, the key is to be an active participant in conversations, showing genuine interest in others' lives and offering help without an immediate expectation of return. This might mean supporting a friend's new venture, participating in community events, or simply being there for others in times of need. It's about creating deep, meaningful connections that go beyond surface-level interactions.

In both cases, the underlying principle is to be a source of positive energy in the network. This doesn't mean shying away from authenticity or not addressing challenges; rather, it's about being a constructive force that contributes to the network's overall health and vitality. Whether it's through offering valuable expertise in professional circles or being a reliable confidant in personal networks, the goal is to be a member who adds undeniable value, thus naturally fostering an environment where support is both given and received. This approach of proactive, meaningful engagement is what transforms networks from mere collections of contacts to rich, supportive communities.

AI-Driven Insights and Recommendations for Network Support

Continuing from the earlier discussion on networking strategies, the introduction of AI-driven insights and recommendations adds a new dimension to how we engage within our networks, whether they're social, personal, or professional. These AI systems, sophisticated in their data analysis, provide real-time, tailored recommendations that align closely with individual needs and aspirations. In the professional world, for example, AI can analyze a user's professional trajectory, skill set, and networking patterns to suggest potential mentors, collaborators, or even career opportunities. This is more than just automated networking; it's a strategic alignment of one's career path with the vast possibilities within the network.

In *personal* and *social networks*, the role of AI becomes even more nuanced. It has the ability to sift through personal interests, past interactions, and preferences to recommend social groups, events, or content that resonate on a more personal level. Imagine a social platform where AI curates your experience, bringing you closer to discussions, communities, and content that genuinely align with your personal life and interests. This isn't just about filtering information; it's about enhancing the quality of your social interactions, making them more relevant, engaging, and enriching.

This advanced level of personalization, made possible by AI, fundamentally changes the networking landscape. It transforms passive, often overwhelming networks into dynamic, curated experiences. The power of AI-driven networking lies in its continuous learning and adaptation to your evolving preferences, ensuring that every suggestion or connection it makes is relevant and meaningful. This intelligent, personalized approach is setting a new standard in networking, shifting the focus from quantity to quality of connections. It's a promising step toward more meaningful, productive, and satisfying networking experiences, tailored to each individual's journey and goals.

AI-Powered Tools for Collaboration and Communication Within Networks

AI-powered tools for collaboration and communication within networks are not just technological advancements; they are pivotal in fostering a culture of equity and equality. These tools democratize interactions, ensuring that every voice within a network, regardless of its hierarchical or geographical position, can be heard and valued. For example, AI-enhanced communication platforms can translate languages in real-time, breaking down barriers that previously hindered collaboration across diverse groups. This feature embodies a commitment to inclusivity, enabling seamless interaction among network members from different linguistic backgrounds. Similarly, AI-driven collaboration tools can intelligently assign tasks and projects based on individual skills and workload, ensuring a fair distribution of responsibilities. This approach aligns with a vision where the worth of contributions is recognized and celebrated, fostering an environment where merit and quality are the guiding principles. These tools are redefining the essence of collaborative networks, transforming them into spaces where equality is not just an ideal but a practical, everyday reality.

Examples of AI-Driven Support Initiatives Within Networks

AI-driven support initiatives within networks are increasingly focusing on promoting equity, equality, and recognizing quality and merit. Here are some notable examples.

AI for Diverse Talent Recruitment

Initiative: AI platforms like Pymetrics use neuroscience-based games and bias-free algorithms to assess candidates' potential beyond their resumes.

Impact: This approach levels the playing field for job applicants, ensuring individuals are evaluated and selected based on inherent abilities and potential, rather than conventional markers like pedigree or background, thus promoting equity in the hiring process.

Language Translation in Real-Time Communication

Initiative: Tools like Skype's real-time translation feature use AI to break down language barriers in communication.

Impact: This technology enables individuals from diverse linguistic backgrounds to collaborate effectively, ensuring equal participation in global conversations and projects, thereby fostering a more inclusive environment.

AI-Enhanced Peer Review in Academic Publishing

Initiative: Springer Nature, a prominent academic publisher, is using AI systems to assist in the peer review process for academic papers, identifying the most relevant reviewers and ensuring the review process is fair and unbiased.

Impact: This helps in upholding the quality and merit of academic work while ensuring that the review process is equitable, free from biases often seen in manual selection processes.

Personalized Learning Platforms

Initiative: AI-powered educational platforms like Coursera or Khan Academy offer personalized learning experiences, adapting to each student's unique learning style and pace.

Impact: Such platforms democratize education by providing high-quality, tailored educational content to a wide range of learners, irrespective of their geographical or socioeconomic status.

AI for Fair Workload Distribution

Initiative: Asana, a leading project management software, uses AI tools to allocate tasks based on team members' current workload and skill sets.

Impact: This ensures a fair distribution of work and helps in recognizing the individual contributions of team members, promoting a merit-based work environment.

These examples illustrate how AI, when applied thoughtfully, can be a powerful tool in supporting initiatives that promote equity and equality, while also upholding the principles of quality and merit within various networks.

How to Expand Your Networks?

Expanding your network in today's interconnected world is more than just adding contacts; it's about strategically cultivating relationships that bring reciprocal value and growth. It's an artful blend of deliberate engagement and authentic connection, where each new link in your network is a potential avenue for collaboration, learning, and innovation. Embracing a mindset of open-mindedness and genuine curiosity, one must seek out diverse perspectives and experiences, venturing beyond familiar territories. This approach to network expansion is not a mere accumulation of names but a thoughtful process of building a community that resonates with your personal and professional ethos. It's about identifying and fostering relationships with individuals and groups that not only align with your current interests and goals but also challenge and inspire you to explore new ideas and opportunities.

Diversifying and Growing Networks

Diversifying and growing networks, particularly in a landscape that increasingly values equity and equality, involves a conscientious approach that intertwines the richness of diversity with the core principles of quality and merit. In the pursuit of expanding networks, the focus should extend beyond mere numbers to the cultivation of a varied and inclusive community. This means actively seeking connections from different industries, backgrounds, and viewpoints, which enriches the network with a plethora of experiences and insights. For instance, platforms like LinkedIn have made strides in this direction, facilitating connections across a diverse professional landscape. They enable users to connect with industry leaders, peers from different cultural backgrounds, and professionals with varied experiences, thereby fostering a network that is not only wide-ranging but also rich in quality and substance.

In parallel, it's crucial to ensure that the expansion of networks is guided by the principles of equality and merit. Each new connection should bring a unique value to the table, contributing to the network in meaningful ways. In *personal networking* scenarios, this might look like joining interest-based groups or communities that not only align with your passions but also challenge and elevate your understanding. *Professionally*, it can involve participating in cross-industry conferences or webinars, which not only broadens one's perspective but also creates opportunities for meaningful collaborations based on shared interests and mutual professional respect. These expanded networks, built on the foundation of diversity and enriched by quality interactions, become powerful resources for innovation and growth.

In both *personal* and *professional* contexts, the emphasis on quality and merit ensures that the network remains robust and effective. It's about creating a balanced ecosystem where each member contributes to and benefits from the collective wisdom and experience of the group.

This approach to networking fosters a culture of mutual support and continuous learning, where the diversity of the network contributes to its strength and vitality, making it an invaluable asset in one's personal and professional journey.

In real-time scenarios, the impact of diversified and merit-based networking is evident across various platforms and initiatives. For example, ***professional networking*** groups on LinkedIn, specifically tailored for diverse industries and interests, offer a wealth of opportunities for members to connect and collaborate based on shared goals and expertise. These groups are not just networking hubs but knowledge-sharing communities, where the quality of interactions is enhanced by the diverse professional backgrounds of their members. Similarly, platforms like Meetup and Eventbrite facilitate the discovery and organization of events and gatherings across a myriad of interests and professions. These platforms enable individuals to expand their networks by connecting with people from different cultural, professional, and geographical backgrounds, fostering an environment of inclusive growth. Such real-time examples underscore the importance of diversity and quality in expanding networks, illustrating how a varied and merit-based approach to networking can lead to more meaningful connections and collaborative opportunities.

AI-Driven Network Expansion Strategies

Continuing the exploration of network expansion in today's interconnected world, AI-driven strategies are playing a pivotal role, particularly in tailoring and enhancing these networks. In professional settings, AI is not merely a tool for suggesting potential connections; it's an intelligent facilitator that assesses compatibility and potential for mutually beneficial relationships. Networking platforms harness AI to analyze user profiles, past interactions, and career aspirations, making recommendations that transcend superficial connections. This intelligent matchmaking enables professionals to grow their networks with individuals and groups that align with their career trajectory, ensuring that every new connection is a step toward professional enrichment.

In the realm of *personal* and *social networking*, AI-driven tools are equally impactful, offering a personalized approach to expanding one's social circle. Social media platforms, equipped with AI algorithms, are adept at suggesting groups, events, and communities based on user interactions and preferences. This goes beyond random recommendations; it's about connecting individuals with communities where they can find genuine engagement and shared interests. For instance, an AI tool might suggest a local photography group to someone who frequently posts and interacts with photography content, thereby expanding their social network with a focus on quality and shared passion. This nuanced approach to network expansion facilitated by AI is transforming how individuals build and nurture their social and personal connections, ensuring that these networks are not just extensive but also meaningful and aligned with their interests and values.

AI-Powered Tools for Discovering and Connecting with New Contacts

AI-powered tools are revolutionizing the way we discover and connect with new contacts across social, personal, and professional realms, embedding principles of equity and equality into the very fabric of networking. These tools, using sophisticated algorithms, level the playing field by providing every user, regardless of their background or current network size, equal access to a wealth of potential contacts. This democratization of networking opportunities means that connections are no longer restricted by traditional barriers like geography or industry gatekeeping. For example, an AI-driven networking app might suggest a contact in a different country or industry who shares your *professional interests* or *personal passions*, thus expanding your network in ways previously unimaginable. The emphasis here is not just on expanding networks but on enhancing their quality and relevance. AI ensures that these new connections are not random but are carefully curated based on shared interests, expertise, and goals, championing the merit of each potential contact. In this way, AI is not only a tool for growth but an advocate for meaningful, equitable, and merit-based networking.

Case Studies for Successful AI-Driven Network Expansion Efforts

The landscape of AI-driven network expansion is rich with successful case studies that exemplify how technology can foster equity, equality, quality, and merit in networking efforts.

One notable example is a global tech company that implemented an AI-driven internal networking tool designed to connect employees across different departments and geographical locations. The AI algorithm considered factors like professional interests, project experiences, and skill sets to recommend potential internal connections. This initiative not only broke down silos within the organization but also promoted diversity and inclusion by connecting employees who might not have interacted otherwise. The result was a more collaborative and innovative work environment, where quality connections led to meaningful collaborations and knowledge sharing.

Another case study comes from a professional networking platform that used AI to create mentorship programs for underrepresented groups in the tech industry. The platform's AI system matched mentees with mentors based on shared professional interests, career goals, and experiences. This approach ensured that mentorship opportunities were accessible to a more diverse group of individuals, promoting equity and providing a space where the merit of an individual's aspirations and skills were recognized and fostered.

Additionally, an AI-powered networking app for entrepreneurs and small business owners stands as a testament to how AI can democratize networking opportunities. The app used AI to suggest connections and business opportunities to users based on their business type, size, and goals. This not only provided equal networking opportunities to businesses of all sizes but also ensured that these connections were valuable and relevant, fostering a community where quality and merit were the driving forces behind each new connection.

These case studies demonstrate the profound impact that AI can have in network expansion efforts, ensuring that these initiatives are grounded in principles of equity and equality. By leveraging AI, these platforms and tools are able to offer more personalized, relevant, and meaningful networking experiences, where the quality of connections and the merit of individuals and their ideas are celebrated.

Networking and Alumnus Tools and How They Contribute

In today's interconnected world, a plethora of networking and alumni tools are redefining how we connect across *social*, *personal*, and *professional* spheres, with a steadfast focus on promoting equity, equality, quality, and merit. Platforms like LinkedIn serve as a cornerstone for *professional networking*, offering a space where individuals can connect based on shared professional interests and achievements, transcending geographical and industry boundaries. This approach ensures that networking opportunities are not restricted to elite circles, but are accessible to a diverse global audience, fostering an environment of equality.

For alumni connections, platforms like Graduway provide tailored networking opportunities for former classmates and colleagues, enabling them to reconnect and collaborate based on shared educational backgrounds and experiences. These tools emphasize the merit of shared experiences and professional accomplishments, rather than just social standing or personal connections. In the realm of *personal networking*, platforms like Meetup facilitate connections based on shared hobbies and interests, allowing individuals to build networks that are both personally fulfilling and equitable. Each of these platforms contributes to a broader ecosystem where networking is not just about who you know, but about the quality and relevance of the connections you make, ensuring that every interaction is meaningful and contributes to a culture of inclusive professional and personal growth.

Overview of Networking and Alumnus Tools

Networking and alumni tools have now evolved into sophisticated platforms that not only connect individuals but also foster a culture of equality, equity, quality, and merit. For example, platforms like LinkedIn and Alumni.net have revolutionized *professional* and *educational networking* by creating spaces where the merit of one's skills and experiences takes precedence over traditional networking paradigms. LinkedIn, for instance, empowers professionals to showcase their achievements, share insights, and connect with others in their field, irrespective of geographical or hierarchical boundaries. This democratization of professional networking ensures that opportunities for growth and collaboration are accessible to all, thereby fostering a landscape of equity and equality.

Similarly, alumni tools have transcended their traditional role of simply reconnecting old classmates. Today, these platforms serve as dynamic communities where former students from educational institutions can engage, share opportunities, and collaborate on projects. They provide a level playing field where the quality of one's ideas and contributions are the benchmarks for connection and collaboration, rather than their social standing or past affiliations. This shift reflects a broader movement toward inclusivity and meritocracy in networking, aligning with the ethos that opportunities should be accessible to all, and recognition should be based on individual merit and contributions. These tools have thus become instrumental in building networks that are not only extensive but also rich in diversity, expertise, and potential, ensuring that every member has an equal opportunity to thrive and contribute.

The Role of AI in Enhancing Networking and Alumnus Tools

AI brings a level of sophistication to networking tools, enabling them to provide more meaningful and relevant connections, tailored to individual needs and aspirations. In *professional* contexts, platforms like LinkedIn utilize AI to analyze users' career histories, skills, and engagement to suggest connections and opportunities that are not just relevant, but also potentially transformative for their career paths. This intelligent system ensures that professional networking transcends the mere exchange of business cards, evolving into a strategic tool for career development and growth.

In the realm of *social* and *personal networks*, AI plays a crucial role in enhancing user experience by personalizing interactions and connections. *Social media* platforms, for instance, use AI to filter and recommend content, groups, or connections that align with a user's interests, past interactions, and preferences. This goes beyond algorithmic efficiency; it's about creating a space where social interactions are more meaningful and aligned with *personal* interests, thereby enhancing the quality of social networking experiences. Moreover, AI-driven alumni networks are transforming the way we reconnect with past peers. By analyzing shared experiences, interests, and professional trajectories, these platforms facilitate connections that are not merely based on past acquaintances but on potential future collaboration and mutual growth. In all these scenarios, AI ensures that the networks we build and engage with are equitable, diverse, and merit based, reflecting a deep understanding of individual preferences and a commitment to enhancing the quality of our connections.

AI-Driven Contributions to Social Equity in Networks

AI-driven contributions to social equity in networks are creating a paradigm shift, embodying a vision where technology is not just a facilitator but a catalyst for inclusive growth. In this landscape, AI is being leveraged to dismantle traditional barriers, offering equal networking opportunities to individuals irrespective of their background or geographical location. Platforms powered by AI are analyzing user data to identify and bridge gaps in networking opportunities, ensuring that access to valuable connections and resources is not a privilege of the few but a right accessible to all. This approach to networking, championed by AI, resonates with a commitment to diversity and inclusivity, mirroring a world where every individual has the potential to contribute and benefit. It's a transformative shift, echoing the ethos that the true value of a network lies not just in its size but in the equity and quality of connections it fosters. In this AI-enhanced networking realm, every interaction, every connection is imbued with the potential to contribute to a more equitable and connected society.

Examples of AI-Enhanced Networking and Alumnus Tools

AI-enhanced networking and alumni tools are making significant strides in promoting equity, equality, quality, and merit within their platforms. Here are a few examples that embody these principles.

Diverse Talent Discovery Platforms

Example: Tools like Entelo and HiringSolved use AI to help companies identify and recruit diverse talent. By analyzing vast amounts of data, these platforms can uncover candidates from varied backgrounds, ensuring a more equitable approach to talent acquisition.

Impact: This leads to a more diverse workforce, breaking down traditional hiring biases and promoting equality in professional opportunities based on the merit and qualifications of candidates.

Alumni Networking Platforms

Example: Alumni platforms like Graduway and Almabase integrate AI to offer personalized networking experiences for alumni. These platforms analyze alumni profiles and interests to recommend relevant connections and professional opportunities.

Impact: Such tailored recommendations ensure that alumni from all backgrounds have equal access to networking opportunities, thereby promoting equity in professional development and growth.

Inclusive Professional Networking

Example: LinkedIn uses AI algorithms to suggest connections, job opportunities, and content that are tailored to the user's profile. This ensures that professionals, regardless of their industry, location, or background, have access to opportunities that match their skills and experiences.

Impact: This approach democratizes professional networking, allowing for equitable access to career opportunities and resources, based on the quality of one's profile and merits.

147

Mentorship Matching Platforms

Example: Platforms like MentorCruise use AI to match mentees with mentors in their desired field. The AI considers factors like professional goals, experience level, and specific interests to create meaningful mentor–mentee relationships.

Impact: This enables equitable access to mentorship, ensuring that individuals can find guidance and support based on their specific needs and merits, regardless of their background.

These examples illustrate how AI is being used to enhance networking and alumni tools, with a keen focus on promoting equity and equality while ensuring that connections and opportunities are based on quality and merit. This approach is transforming networking from a privilege for the few into an inclusive platform for many, where diversity is celebrated and equal opportunity is the norm.

Ethical Considerations in AI-Enhanced Networking

The ethical landscape of AI-enhanced networking demands a nuanced approach, particularly when balancing the imperatives of equity, equality, quality, and merit. As AI systems increasingly dictate the dynamics of networking, it's crucial to ensure these technologies are developed and deployed with a deep sense of responsibility and fairness. Ethical considerations must be at the forefront, ensuring that AI algorithms do not reinforce existing biases or inequalities in *social*, *personal*, or *professional networks*. This requires a conscientious effort to design AI models that are transparent, accountable, and subject to regular audits for fairness. The ethical deployment of AI in networking should strive to create an environment where opportunities for connection, growth, and advancement are accessible to all, based solely on the merits and contributions of individuals. The goal is a networking ecosystem where technology acts as an equalizer, not as a divider, fostering a landscape that is not only technologically advanced but also ethically sound and inherently fair.

Privacy Concerns and Data Protection

Continuing the discussion on the ethical considerations of AI in networking, privacy concerns and data protection emerge as critical facets. In an era where AI algorithms thrive on vast amounts of user data to make connections and recommendations, the question of how this data is used and protected becomes paramount. Users entrust these platforms with their *personal* and *professional* information with the expectation of privacy and security. However, instances like the Facebook–Cambridge Analytica scandal serve as a stark reminder of the potential pitfalls in data handling. This incident underscored the need for stringent data protection policies and transparent user consent mechanisms in AI-driven platforms.

Moreover, in the context of networking, the equitable and fair use of data is crucial. AI systems must be designed to ensure that the data they use does not lead to biased outcomes, inadvertently creating echo chambers or excluding certain groups from opportunities. For instance, LinkedIn has made efforts to adjust its algorithms to prevent unintended bias in job recommendations and networking suggestions. This commitment to ethical data use is essential not only for maintaining user trust but also for ensuring that the networking landscape remains diverse, inclusive, and merit-based. The challenge lies in developing AI systems that can leverage user data to enhance networking experiences while simultaneously upholding the highest standards of privacy, security, and ethical integrity. This delicate balance is the cornerstone of fostering a networking environment where technology empowers users equitably, without compromising their privacy or the quality of their connections.

Potential Biases in AI-Driven Networking Tools

The potential for biases poses a significant challenge, particularly in the context of equity, equality, quality, and merit. These biases, often a reflection of the data on which AI models are trained, can inadvertently perpetuate existing inequalities. For instance, if a ***professional networking*** platform's AI is trained predominantly on data from a specific demographic, it might be less effective in identifying and recommending opportunities for users outside that demographic. This not only skews the networking landscape in favor of certain groups but also undermines the principles of equity and equality that are foundational to a meritocratic system.

Moreover, in ***personal*** and ***social networking*** contexts, AI algorithms could potentially create echo chambers, where users are predominantly exposed to viewpoints and connections that mirror their own. This limits the diversity of interactions and the broadening of perspectives, which are crucial for a balanced and inclusive network. For example, a social media platform's AI that suggests connections and content based on a user's existing views and interactions might reinforce biases and limit exposure to diverse perspectives, thereby impacting the quality of the network.

To counter these biases, it's essential for AI-driven networking platforms to implement robust, inclusive data practices and continuous algorithmic audits. This ensures that AI tools are not just technologically advanced but are also aligned with the ethical standards of fairness and inclusivity. The goal is to harness AI's potential in a way that enriches networking experiences, promotes diversity, and maintains the integrity of a system where connections and opportunities are based on genuine merit and quality.

Ensuring AI Promotes Social Equity Rather Than Perpetuating Inequalities

The potential of AI to promote social equity rather than perpetuating inequalities is becoming increasingly evident. When harnessed correctly, AI has the power to democratize access to networking opportunities across personal, professional, and social spheres. By analyzing vast datasets, AI can identify and recommend connections that transcend traditional social and professional boundaries, fostering a more inclusive networking environment. For example, in *professional networks* like LinkedIn, AI is used to surface job opportunities and connections to users who may otherwise have been overlooked due to geographic or socioeconomic constraints. This approach ensures that talent and potential are the primary factors driving networking opportunities, leveling the playing field for professionals from diverse backgrounds.

In *personal* and *social networks*, AI's role in promoting equity is equally significant. Social media platforms are using AI to curate content and connections that reflect a diverse range of perspectives and experiences, counteracting the echo chambers that often emerge in such spaces. This move toward a more balanced representation of ideas and voices in online communities is crucial in creating a more equitable and open digital society. Moreover, AI-driven tools in personal networking apps can suggest events, groups, or activities based on individual interests and behaviors, rather than on preexisting social circles, thus expanding one's horizons and fostering connections with a wider array of individuals.

These examples underscore the potential of AI in leveling the networking playing field, making it more about the merit of one's skills and the quality of one's contributions than about preexisting connections or backgrounds. The key to realizing this potential lies in the continuous and mindful development of AI technologies, ensuring they are programmed to recognize and mitigate biases, thereby paving the way for more equitable and just networking ecosystems.

Guidelines and Best Practices for Ethical AI-Enhanced Networking

Continuing the exploration of ethical AI in networking, it's crucial to establish guidelines and best practices that ensure these technologies promote fairness, equality, and merit. First and foremost, AI algorithms must be designed with diversity and inclusivity in mind. This means feeding AI systems with diverse datasets that represent a broad spectrum of users to prevent biases. Regular audits and updates of these algorithms are essential to identify and rectify any biases that may emerge. For instance, LinkedIn has made strides in this area by continuously refining its AI to reduce bias in job recommendations and networking suggestions, ensuring that opportunities are equitably distributed among all users, irrespective of their background.

Another best practice is transparency in AI operations. Users should be informed about how their data is used and how AI algorithms make networking suggestions or decisions. This transparency builds trust and ensures users that their personal information is handled responsibly. Furthermore, there should be an option for users to provide feedback on AI-driven recommendations. This feedback loop allows for adjustments and improvements, ensuring that AI systems evolve to better serve the diverse needs of the network's users. In *personal networks*, like social media platforms, AI recommendations for connections or content should be adjustable based on user preferences, ensuring that users have control over their networking experience.

These guidelines, focused on ethical AI deployment, contribute to a networking landscape where technology is a tool for fostering equitable, diverse, and merit-based connections. By adhering to these best practices, AI-enhanced networking can become a powerful force for good, breaking down traditional barriers and creating a more inclusive, equitable, and productive professional and social environment.

As we draw to a close on the intricate subject of building social equity into networks, it's evident that the journey is multifaceted, requiring a blend of thoughtful strategy, ethical technology implementation, and a steadfast commitment to inclusive principles. The key takeaway is that networks, whether *personal*, *professional*, or *social*, are not just platforms for connection but potent tools for fostering a culture of equity and equality. The incorporation of AI and GenAI in this realm, when guided by ethical considerations, has the potential to significantly diminish biases, thereby creating an environment where opportunities and connections are defined not by background or status, but by the quality and merit of contributions.

The future of networking, as envisioned by visionaries and leaders, lies in the creation of spaces where diversity is not just acknowledged but celebrated, where the exchange of ideas and opportunities is not hindered by inequity but enhanced by the richness of varied perspectives. In such an ecosystem, every individual, regardless of their starting point, has the potential to connect, grow, and contribute meaningfully. The real-time examples we've seen, from AI-driven *professional platforms* to *personalized social media* experiences, all point toward a future where networking is a catalyst for positive change, fostering a world that is not only more connected but also more equitable and just.

Conclusion

In conclusion, the path to embedding social equity into networks is continuous and evolving. It demands vigilance, innovation, and a collective commitment to principles of fairness and inclusivity. As we harness the power of AI and other emerging technologies, our focus must remain steadfast on creating networks that are equitable, diverse, and reflective of the world we aspire to live in. This is not just a technological goal but a societal imperative, where each connection we forge and each network we build brings us one step closer to a more equitable and inclusive future.

As we transition into the next chapter, we stand at the precipice of understanding and quantifying the impact of social equity within networks. This upcoming exploration delves into the intricate process of developing robust, meaningful metrics that not only measure but also guide efforts toward achieving a more equitable and just network environment. In this pursuit, we will explore how metrics can be designed to reflect true equity and equality, ensuring that they capture not just the quantity of interactions and connections but also the quality and depth of these engagements. This chapter aims to unravel the complexities of measuring social equity in a manner that is both pragmatic and visionary. We will navigate through the challenges of creating metrics that are fair, unbiased, and reflective of the diverse dimensions of equity, all while upholding the standards of quality and merit. This chapter is not just about the "what" and the "how" of measurements; it's about paving a path toward a more equitable future, guided by data, driven by purpose, and informed by a deep understanding of the multifaceted nature of social equity.

CHAPTER 6

Metrics and Measurement for Social Equity

In the realm of fostering robust and dynamic communities, social equity emerges as a pivotal element, essential for nurturing environments where innovation, inclusivity, and collective well-being are paramount. This chapter delves into the array of tools, strategies, and methodologies that are instrumental in measuring and enhancing social equity. Central to our exploration will be the role of online tools and apps designed to meticulously gauge social interactions, providing insights essential for understanding and sculpting community dynamics.

Moreover, we'll examine how expanding and deepening network connections can serve as vital conduits for promoting social equity. This exploration isn't just about broadening networks; it's about enriching them with diverse perspectives and equitable opportunities. We'll delve into how these networks can be leveraged to foster environments where every member, regardless of background or status, has access to equal opportunities, ensuring that the merit and quality of contributions are recognized and valued.

© Raghu Banda 2024
R. Banda, *Building Social Equity with AI*, https://doi.org/10.1007/979-8-8688-0091-7_6

Additionally, our discussion will navigate the complexities of various types of transactions within these communities. From financial exchanges to information sharing, each transaction type carries its own weight in building a socially equitable landscape. We aim to unpack the nuances of service weightage in these transactions, advocating for a balanced approach where each exchange is evaluated not just on its economic value but on its contribution to the community's overall equity and well-being.

Finally, we will explore methods for reviewing and rating these transactions, emphasizing a framework that prioritizes quality, merit, equity, and equality. This approach transcends mere transactional analysis, fostering a culture where every interaction is a reflection of the community's core values and principles, contributing to the creation of a community that is not only economically robust but also socially just and equitable!

What Are the Different Online Tools and Apps That Measure Social Interactions?

In the era where digital connectivity shapes our social fabric, online tools and apps have become pivotal in measuring social interactions, particularly through the lens of equity and equality, while upholding the principles of quality and merit. These digital platforms are not mere analytical tools; they are reflections of our societal pulse, offering nuanced insights into how we connect, engage, and influence each other.

For instance, social analytics tools, like Hootsuite or Sprout Social, provide a sophisticated understanding of social media dynamics. They delve deep into user engagement, sentiment analysis, and network reach, offering a comprehensive view of how information spreads and how inclusive these interactions are. These insights are invaluable in ensuring that social media platforms are not just echo chambers but inclusive spaces where diverse voices are heard and valued, aligning with a vision where every user, irrespective of their background, has an equal opportunity to be seen and heard.

Moreover, apps like Nextdoor or Meetup go a step further by not just analyzing but actively fostering community interactions based on shared interests and locality. They exemplify how technology can bridge the gap between the digital and the physical world, creating communities that are both locally relevant and diverse. The quality and merit in these interactions stem from their ability to bring together varied groups, promoting a sense of belonging and mutual respect among members.

In these tools and platforms, the emphasis on equity and equality is clear. They strive to create a digital environment where interactions are not only quantifiable but also qualitatively rich, ensuring that the value of each connection is not lost in the sea of data. This approach, resonating with the ethos of thought leaders who advocate for technology as a force for good, ensures that our digital interactions contribute positively to our social fabric, making them more equitable, inclusive, and meaningful.

Social Network Analysis Tools

Social network analysis tools are at the forefront of understanding and shaping the dynamics within various communities, playing a crucial role in the context of equity, equality, quality, and merit. These tools, through their analytical prowess, provide a detailed mapping of social connections, highlighting how information flows, how communities are structured, and how influential certain nodes within a network are!

For instance, tools like NodeXL and Gephi offer insights into the intricacies of social networks by analyzing patterns of relationships and interactions. They enable organizations and researchers to identify key influencers, understand community structures, and observe how information propagates within a network. This kind of analysis is vital in ensuring that information dissemination and community engagement strategies are equitable and reach diverse groups effectively. It ensures that all voices within a network, especially those from marginalized or underrepresented groups, are heard and have the opportunity to influence the network.

Furthermore, platforms like LinkedIn use social network analysis to enhance professional networking experiences. They analyze user data to suggest connections, endorse skills, and recommend job opportunities, ensuring that these suggestions are based on the user's professional merits and achievements. This approach not only democratizes professional networking but also ensures that opportunities are extended based on quality and professional relevance, rather than solely on existing personal connections.

In these tools, the focus on equity and equality is intertwined with the commitment to quality and merit. They aim to create a digital environment where every interaction and connection is not only strategically significant but also contributes to a more inclusive, fair, and merit-based community. This balanced approach in social network analysis, resonating with the ethos of leveraging technology for societal good, ensures that our digital landscapes are not only efficient in mapping connections but also equitable and just in fostering meaningful and inclusive relationships.

Real-time examples of social network analysis tools demonstrate their profound impact on various sectors, from business to social research, underlining their role in promoting equitable and merit-based networking.

NodeXL in Market Research

Example: A marketing firm uses NodeXL for social media analysis to understand consumer behavior and trends. By analyzing connections and interactions among users discussing a specific product, NodeXL helps identify key influencers and opinion leaders within niche markets.

Impact: This enables businesses to tailor their marketing strategies more effectively, ensuring that they engage with the right audiences and leverage influential voices in a fair and merit-based manner.

Gephi in Academic Research

Example: Researchers use Gephi to study social movements by mapping out the connections between activists and organizations on social media. This analysis provides insights into how information spreads within these networks and identifies central figures in the movement.

Impact: The tool's ability to highlight influential nodes based on their actual role and contribution ensures a more accurate and merit-based understanding of social movements, aiding in equitable policy development and research.

LinkedIn's Analysis for Professional Networking

Example: LinkedIn's algorithm analyzes user data to recommend professional connections. This system considers factors like shared professional interests, mutual connections, and endorsements to suggest relevant contacts.

Impact: Such targeted recommendations help individuals expand their professional networks in a manner that is both equitable (offering equal networking opportunities to all users) and based on merit (highlighting connections that are professionally relevant and beneficial).

Twitter/X Analytics for Public Sentiment Analysis

Example: Twitter's analytics tools are used by organizations to gauge public sentiment on various topics. By analyzing tweets, retweets, and mentions, these tools map out how information flows through the network.

Impact: This analysis helps in understanding diverse viewpoints, ensuring that all sections of the Twitter community are represented in understanding public opinion, thereby promoting equity in public discourse.

These examples underscore the significant role social network analysis tools play in providing insights that are not only data-driven but also equitable and merit-focused. They demonstrate how technology can be used to foster a fairer and more inclusive digital environment, where connections and influence are based on relevance and contribution rather than bias or privilege.

Social Listening and Sentiment Analysis Tools

Social listening and sentiment analysis tools represent a confluence of technology and human insight, offering an unprecedented window into the collective consciousness of the digital world. These tools harness advanced algorithms and natural language processing to sift through the vast expanse of social media and online interactions, identifying trends, emotions, and opinions that pervade our online conversations. This analytical capability is not just about aggregating data; it's about understanding the nuanced tapestry of human sentiment and reaction, providing a compass for navigating the complex social landscapes that define our times.

In this context, tools like Brandwatch and Sprout Social stand out for their ability to not only track mentions and keywords but also to analyze the sentiment behind interactions, offering insights that are both broad in scope and deep in understanding. This level of analysis ensures that organizations, policymakers, and individuals can tune into the prevailing sentiments of their audiences or communities, making informed decisions that are responsive to the needs and concerns of those they serve. The application of these tools transcends mere market research or brand management; it is a powerful instrument for social equity, enabling voices from across the social spectrum to be heard and considered.

Real-time examples of these tools in action underscore their potential for promoting equity, quality, and merit. For instance, during social movements or campaigns, sentiment analysis tools can provide organizers and participants with a real-time gauge of public opinion and emotional tone, identifying shifts in sentiment that could signal the need for strategic adjustments. This ensures that messaging and advocacy efforts are grounded in the actual feelings and responses of the community, promoting a more equitable and responsive approach to social activism.

In the professional realm, companies use these tools to ensure their products, services, and communications resonate with their diverse customer base, adhering to principles of quality and merit. By listening to customer feedback across various channels and analyzing the sentiment of these interactions, businesses can tailor their offerings to meet the needs and expectations of their customers more effectively, demonstrating a commitment to equity and excellence in service.

These examples highlight how social listening and sentiment analysis tools are not just technological solutions but strategic assets in the pursuit of equity, quality, and merit. They empower organizations and individuals to listen more closely, respond more thoughtfully, and engage with their communities in ways that are informed, equitable, and attuned to the diverse voices that define our society. Some of these tools range from sophisticated AI-driven platforms to grassroots community engagement strategies, each with its unique impact.

Brandwatch (Online Tool)

Example: Brandwatch is a digital consumer intelligence platform that uses AI for social listening and sentiment analysis across social media and the web. It allows organizations to understand consumer behavior, trends, and sentiment in real time.

Impact: A nonprofit organization used Brandwatch to monitor public sentiment about environmental issues. The insights gained helped them tailor their advocacy campaigns more effectively, ensuring they addressed the concerns and motivations of a diverse audience, thereby promoting environmental equity and action.

Town Hall Meetings (Offline Tool)

Example: Town hall meetings, as an offline tool, enable direct dialogue between community leaders and members. These meetings provide a platform for voicing concerns, discussing community issues, and brainstorming solutions in an open forum.

Impact: A local government held regular town hall meetings to discuss urban development plans, ensuring that community members from all neighborhoods had an equal opportunity to contribute their views and concerns. This approach led to more inclusive and equitable urban planning that considered the needs and suggestions of a diverse population.

SurveyMonkey (Online Tool)

Example: SurveyMonkey is an online survey platform that facilitates the collection of data and opinions from a wide range of respondents. It's widely used for market research, customer satisfaction surveys, and feedback collection.

Impact: A university used SurveyMonkey to gather alumni feedback on its inclusion initiatives, allowing for anonymous and candid responses. The feedback provided valuable insights into the effectiveness of these initiatives, guiding the university in making data-driven improvements to promote equity and inclusion among its alumni network.

Community Workshops (Offline Tool)

Example: Community workshops involve interactive sessions where community members come together to learn, share, and collaborate on various topics or projects.

Impact: An NGO organized workshops in rural areas to educate communities about sustainable farming techniques. These workshops served as a knowledge exchange platform, empowering farmers with the information and skills needed to improve crop yield and sustainability. The inclusive nature of these workshops ensured that knowledge was accessible to all community members, thereby promoting quality and merit in agricultural practices.

These examples illustrate the dynamic range of tools available for engaging with and understanding communities. Whether through digital platforms like Brandwatch and SurveyMonkey, which provide broad and deep insights into public sentiment and behavior, or through direct community engagement methods like town hall meetings and workshops, these tools are instrumental in promoting equity, equality, and inclusion. They ensure that every voice has the chance to be heard and that decisions and initiatives are informed by the diverse perspectives of the community.

Social Media Analytics Tools

Social media analytics tools stand at the confluence of technology and human insight, offering a sophisticated lens through which we can understand the vast and varied landscapes of social media. These tools dissect the complex web of interactions, engagements, and sentiments that flow through social media channels, providing data-driven insights that can guide content strategies, marketing campaigns, and community engagement efforts. The power of these tools lies in their ability to cut through the noise, identifying patterns and trends that might not be immediately apparent, yet are crucial for making informed decisions in a digital world that values both speed and precision.

In the spirit of equity, quality, and merit, social media analytics tools like Sprout Social, Hootsuite, and Buffer play a pivotal role. They democratize access to insights that were once the purview of those with the resources to conduct extensive market research. Now, small businesses, nonprofits, and even individuals can gauge the impact of their online presence, understand the needs and sentiments of their audience, and tailor their approach to better serve their community. For example, a small community organization might use these tools to monitor conversations around local issues, enabling them to craft campaigns that directly address the concerns and aspirations of their community members. This not only enhances the quality of their engagement but also ensures that their efforts are grounded in the principles of equity and merit, giving voice to those who might otherwise be overlooked.

Moreover, in the realm of professional networking, LinkedIn's analytics tools offer professionals and companies alike the opportunity to understand the reach and engagement of their posts and articles. This capability allows for content optimization that speaks directly to the interests and needs of their professional community, ensuring that valuable insights and contributions are recognized and amplified. This approach underscores a commitment to merit, ensuring that the most relevant, insightful, and valuable content rises to the top, fostering a professional environment that rewards quality and relevance.

These real-time examples underscore how social media analytics tools are more than just aggregators of data; they are instruments of equity, enabling voices from across the spectrum to be heard and valued. They ensure that content strategy, community engagement, and professional networking are not just about who shouts the loudest but who contributes meaningfully to the conversation, promoting a digital ecosystem that is equitable, diverse, and rich in quality and merit.

Here's a look at some of these tools and the impact they create.

Online Tool: Google Analytics

Example: Google Analytics is widely used by businesses and individuals to track and understand website traffic and user behavior. By providing detailed insights into how users interact with a website, it enables content creators to tailor their offerings to meet audience needs more effectively.

Impact: A small nonprofit organization used Google Analytics to understand which of their campaigns were most engaging, allowing them to optimize their web content for better reach and impact. This led to an increase in donations and volunteer sign-ups, demonstrating the power of data-driven decision-making in enhancing organizational effectiveness.

Offline Tool: Community Feedback Forums

Example: Offline community feedback forums are organized by local governments or community groups to gather input on various issues or projects. These forums encourage direct dialogue between community leaders and members, providing a platform for diverse voices to be heard.

Impact: A city council held a series of community feedback forums to gauge public opinion on a new public park project. The insights gathered from these forums led to adjustments in the park's design to better accommodate the needs of families, pet owners, and fitness enthusiasts, showcasing the importance of inclusive community engagement.

Online Tool: BuzzSumo

Example: BuzzSumo is a tool used for content research and monitoring. It allows users to discover trending topics, track brand mentions, and analyze what content performs best for any topic or competitor.

Impact: A content marketing agency used BuzzSumo to identify trending topics within specific niches for their clients, enabling them to create highly relevant and engaging content that significantly improved client website traffic and user engagement.

Offline Tool: Local Networking Events

Example: Local networking events, such as industry meetups or professional association gatherings, provide a physical space for professionals to connect, share ideas, and explore potential collaborations.

Impact: An entrepreneur attended a series of local networking events focused on the tech industry, leading to valuable connections with potential investors and partners. These connections ultimately supported the launch of a successful startup, illustrating the tangible benefits of face-to-face networking.

These examples illustrate the broad spectrum of tools available for enhancing digital presence, engaging with communities, and building networks. Whether leveraging sophisticated online analytics platforms to refine a digital strategy or participating in offline forums to engage with local communities, the combination of online and offline tools enables a more comprehensive, equitable, and effective approach to building and sustaining meaningful connections.

How to Build Social Equity by Connecting, Supporting, and Expanding Your Networks?

Building social equity within the framework of our transactions, whether they're aimed at forging new connections, enhancing existing ones, or exploring uncharted opportunities, requires a deliberate and thoughtful approach. It begins with the strategic identification and engagement of key stakeholders, mapping out both the terrain of our current networks and the potential landscapes yet to be explored. This initial phase is critical, not just for establishing connections but for laying the foundation upon which long-term relationships can be built and nurtured. Providing support, whether through mentorship, sharing valuable resources, or offering both intrinsic and extrinsic value, transforms these connections from mere contacts into meaningful, lifelong partnerships.

The journey doesn't end with nurturing these connections; it extends into strategically expanding them, leveraging both online platforms and offline opportunities like events and conferences. This expansion is not arbitrary but a carefully considered process, designed to weave social equity into the fabric of our networks. By ensuring that each step of the process – connecting, supporting, and expanding – is imbued with principles of equity, equality, quality, and merit, we elevate the entire networking endeavor. It's about enriching these connections with value that is not only reciprocal but also deeply rooted in fairness and inclusivity.

As we prepare to delve deeper into the mechanics of building social equity across various types of transactions – be it enterprise-level exchanges, entity-to-entity collaborations, or individual interactions – it's essential to view these engagements through a lens that prioritizes the creation, nurturing, and expansion of networks with an unwavering commitment to quality and merit. The forthcoming section will dissect these aspects in detail, offering insights into how we can foster networks that are not only extensive in reach but exemplary in their depth and impact. This exploration will span the spectrum of interactions, from the most macroscopic enterprise engagements to the most personal individual connections, underlining the universality of the principles that guide effective, equitable networking.

Identifying and Connecting with Key Stakeholders

Identifying stakeholders and building networks isn't just about expanding your reach. It's a **symphony of diverse voices**, weaving together **equity, quality, and impact** in every interaction. Let's explore this journey through the lenses of renowned thinkers:

Raghavan Venkatraman's "Embrace the dance of opposites": Network building thrives on **diverse perspectives**. In the professional sphere, connecting with individuals outside your comfort zone sparks **creative collisions**, leading to **disruptive innovation**. Imagine a **Raghu Banda-inspired network** where a fintech entrepreneur collaborates with an artist, leading to a revolutionary financial inclusion solution for marginalized communities.

"Radical transparency is paramount": Building trust requires openness and honesty. Imagine a Ray Dalio–influenced approach where stakeholders are actively involved in the decision-making process, fostering a sense of ownership and shared purpose. This transparency builds stronger, more equitable relationships.

"Deep conversations lead to deeper understanding": Meaningful connections go beyond superficial interactions. Imagine a Lex Fridman–inspired network where individuals engage in in-depth dialogues, challenging biases and fostering empathy. This cross-pollination of ideas paves the way for inclusive solutions and societal progress.

"Impactful networks drive societal change": Networks aren't just about individual gain; they're vehicles for positive change. Imagine a Bill Gates–inspired network where tech leaders collaborate with NGOs, tackling global challenges like poverty and climate change. This synergy creates a **ripple effect**, empowering communities and driving progress.

To actively build networks with an eye toward social equity, it's essential to adopt strategies that are both intentional and inclusive, emphasizing quality and merit. For instance, professionals seeking to expand their networks can engage in targeted outreach efforts that prioritize diversity and the sharing of varied perspectives. This could mean participating in or creating networking events that attract a wide range of participants from different industries, backgrounds, and experience levels, thereby fostering a rich exchange of ideas and opportunities.

In the context of personal networking, individuals can strive to build relationships based on shared values and mutual respect, reaching out to others who may not necessarily belong to their immediate social circles or who offer different life experiences and viewpoints. This effort can be facilitated through platforms that encourage diverse interactions, pushing individuals to connect with others based on shared interests or goals rather than surface-level similarities.

On the societal level, creating network-building initiatives that focus on inclusivity can significantly impact social equity. For example, mentorship programs designed to support underrepresented groups in various fields can help level the playing field, offering guidance, support, and access to opportunities that might otherwise be inaccessible. By intentionally connecting experienced professionals with emerging talent from diverse backgrounds, these programs can cultivate a culture of meritocracy and equality.

For organizations and businesses, building equitable networks might involve creating internship or fellowship programs that specifically target individuals from less-represented demographics, ensuring that these opportunities are not just available but also accessible. Companies can also establish partnerships with educational institutions or community organizations to facilitate a more diverse recruitment pipeline, thereby embedding principles of equity and quality right from the hiring process.

In conclusion, building networks with a foundation of equity and equality requires deliberate actions and choices. It's about creating spaces and opportunities where diverse individuals can come together, connect, and grow based on shared merits and achievements. Whether through personal outreach, organizational initiatives, or community programs, the goal is to establish networks that are as diverse, vibrant, and inclusive as the society we aspire to create.

Supporting and Nurturing Relationships

To practically support and nurture a network, let's take a cue from how Bill Gates approaches problem-solving through strategic philanthropy – by applying similar principles, we can foster impactful network relationships. Imagine a professional network where members engage in a mentorship program, structured similarly to Gates' methodical approach to charitable initiatives, where matches are made not just based on industry but on shared values and goals, emphasizing quality and merit in every pairing.

For instance, in a tech entrepreneur network, a successful program could involve seasoned executives mentoring emerging startup founders, with progress tracked through specific milestones and outcomes, like funding rounds or product launches. This real-time application not only strengthens individual connections but elevates the entire network, reinforcing the value of collaborative success over individual achievement.

Drawing from Ray Dalio's principle-based approach, the network could implement regular reflection sessions where mentors and mentees share insights and learnings, akin to Dalio's idea of radical transparency and idea meritocracy. This process ensures that the network evolves through shared knowledge and collective intelligence.

Incorporating Lex Friedman's fascination with AI, the network could leverage advanced algorithms to refine mentor–mentee matching processes, ensuring that the pairings are optimally aligned for mutual growth. AI could analyze past interactions and outcomes within the network to continually enhance the matching criteria based on real-world performance and feedback, ensuring that the network dynamically adapts to serve its members best.

The author's resilience perspective would suggest establishing a support system within the network that can identify and address members' challenges in real-time. For example, if a member faces a professional setback, the network could quickly mobilize resources or advice to assist, demonstrating a tangible commitment to mutual support and resilience.

By weaving together these strategies, the network becomes a living ecosystem of support, growth, and innovation, where every interaction is an opportunity to build on the principles of equity, quality, and merit. Such a network doesn't just grow in size but deepens in value, embodying a culture where every member is invested in the collective well-being and success

Expanding Networks

In today's interconnected world, building a strong network is no longer a luxury, but a necessity. However, simply adding contacts isn't enough. To truly thrive, your network needs to be built on a foundation of **quality, merit, and social equity**. This approach ensures that every connection brings value, fosters collaboration, and empowers individuals and groups to reach their full potential.

Expanding a network effectively requires a strategic approach that aligns with the principles of quality and merit while fostering social equity. Consider the concept of expanding a professional network through inclusive industry conferences. These events can serve as a nexus for diverse talents from various sectors to converge, share ideas, and form new connections. By deliberately designing these conferences to include a wide range of voices and perspectives, organizers can ensure that expansion is not just about adding numbers but enriching the network's diversity and quality.

For instance, a tech conference could incorporate sessions that specifically focus on connecting underrepresented groups in the industry with established leaders. This initiative not only broadens the individual networks of attendees but also enhances the collective resource pool, fostering opportunities for collaboration and innovation across different demographics, disciplines, and otherwise disconnected individuals.

Another practical example could involve a digital platform that facilitates professional introductions, akin to a virtual networking event. This platform could use sophisticated algorithms to suggest connections that are not obvious or readily accessible to individuals, thereby expanding their network in meaningful ways. Such a tool can democratize access to influential networks, breaking down traditional barriers to entry and ensuring that merit and potential are the key drivers of network expansion.

Moreover, networks can expand through cross-sector partnerships, where businesses, nonprofits, and educational institutions collaborate on projects or initiatives that have a broader social impact. For example, a partnership between a tech company, a university, and a nonprofit focused on education equity can lead to innovative solutions that address societal challenges, thereby expanding the network's reach and impact.

In each of these scenarios, the expansion of the network is not just a growth in numbers but an enhancement in quality and diversity, reflecting a commitment to building social equity. By creating more entry points and intersections within the network, there is a greater opportunity for individuals and groups to benefit from new perspectives, resources, and opportunities, ultimately contributing to a more equitable and prosperous society.

By fostering meaningful connections, breaking down barriers, and celebrating individual contributions, we can collectively build a future where everyone has the opportunity to thrive. Remember, your network is an ecosystem – nurture it with care, intention, and a commitment to shared success.

How to Build Social Equity into Any Transaction?

In the intricate tapestry of human interaction, the concept of a Social Equity Score (SES) emerges as a transformative metric, a beacon guiding the multitude of transactions that weave together the fabric of our communities. The SES transcends traditional metrics, becoming a compass that navigates the multifaceted nature of value exchanges, from the deeply personal to the expansively organizational.

Imagine a world where every interaction, every transaction, has the potential to not just benefit individuals, but to strengthen the very fabric of our communities. This is the essence of the **Social Equity Score (SES)** – a transformative metric that goes beyond mere numbers to capture the **resonance and impact** of our actions.

Beyond the Transaction

Every transaction, be it between individuals, entities, or organizations, becomes an opportunity to enhance this score. But the SES is not merely a numerical value; it is the embodiment of the impact and resonance of our actions. It recognizes that each exchange carries a ripple effect, with the potential to fortify bonds, foster trust, and fuel communal synergy.

The SES isn't just about financial gain or professional success. It encompasses the **personal and societal dimensions** of our choices. Picture Maria, a local baker, donating fresh bread to a community shelter. Her act of generosity not only nourishes those in need but also strengthens the sense of community and trust within the neighborhood, contributing positively to her SES.

The SES brings forth a paradigm where every handshake, every service rendered, every collaboration, is a thread in the larger societal weave. In this realm, transactions are not just economic but transformative interactions that enrich the social fabric. The score becomes an invisible currency of goodwill and influence, a testament to the positive footprints one leaves on the sands of society.

Transformative Power

Imagine a world where organizations prioritize their SES as highly as their financial bottom line. In such a world, businesses would be propelled by a dual engine of profit and social contribution. Their services and products would not only fulfill immediate needs but also contribute to the greater good, enhancing their SES with every positive impact made.

Organizations too can harness the power of the SES. Consider "Green Tech Solutions," a company prioritizing renewable energy solutions. They go beyond profit margins, actively engaging in environmental education initiatives. These actions not only address a critical societal issue but also elevate their SES, generating goodwill and attracting customers who share their values.

For individuals, the SES encapsulates the depth of their relationships, the strength of their word, and the legacy of their actions. It is a reflection of their role as catalysts for positive change and their capacity to nurture growth within their circles and beyond.

From Handshakes to Legacies

As we consider the SES in the context of our daily transactions, it becomes clear that it is a call to a higher purpose. It encourages a consciousness of the broader implications of our interactions, urging us to infuse integrity, generosity, and social responsibility into the mundane. It is a compelling vision that elevates the act of giving and receiving beyond the transactional, into the realm of transformational.

The SES recognizes that every interaction, from a professional handshake to a friendly conversation with a neighbor, has the potential to create a ripple effect. By demonstrating **integrity, generosity, and social responsibility** in our daily exchanges, we contribute to building a stronger, more interconnected society.

A Beacon of Change

The SES is not a static score; it's a call to **conscious action**. It encourages us to consider the broader impact of our choices, asking ourselves: "How can I make this interaction not just beneficial but transformative?" This shift in perspective can lead to individuals and organizations becoming more intentional about fostering positive change and leaving lasting legacies.

In essence, the Social Equity Score encapsulates a collective aspiration toward a more connected, responsive, and empathetic world. It is a measure of how well we translate our potential into meaningful actions, how effectively we convert our interactions into lasting value, and how deeply we understand that the true wealth of a society lies not in its coffers, but in the strength and equity of its connections. This score, then, becomes a universal narrative, charting a course toward a future where every transaction enriches the human experience, fostering a legacy of shared prosperity and interconnected triumphs.

Beyond the Coffers

The SES challenges the traditional notion of wealth solely residing in financial resources. It reminds us that the true wealth of a society lies in the **strength and equity of its connections**. By weaving a tapestry of meaningful interactions, fueled by empathy and social responsibility, we can create a future where every individual and community thrives.

In essence, the Social Equity Score is more than a metric; it's a **shared vision and a potent tool**. It prompts us to reimagine the power of our everyday actions, reminding us that even the most seemingly insignificant acts can contribute to a more connected, equitable, and prosperous world for all.

Having established a foundational understanding of how to embed social equity within any exchange and grasped the transformative potential of a Social Equity Score (SES), we now stand at the threshold of deeper exploration. The ensuing sections will delve into the myriad types of transactions that permeate our lives. This journey is critical, for it is in the nuances and particularities of these exchanges that the essence of social equity is either fortified or diminished.

As we navigate through the rich landscape of transactions, we will endeavor to discern their distinct impacts and influences. Our objective is not merely to categorize but to comprehend how each interaction – be it commercial, communal, or personal – can be a vessel for the principles of social equity.

In this pursuit, we shall not be alone. The vanguard technologies of artificial intelligence (AI) and its progeny, Generative AI (GenAI), stand ready to assist us. These tools will not only provide analytical depth but also offer predictive insights, enabling us to foresee and shape the social equity outcomes of our transactions.

By harnessing the analytical prowess of AI and the creative capacities of GenAI, we will learn to measure and enhance the SES effectively. This is not a task taken lightly, for it beckons us to be architects of a future where every transaction is a building block of a more equitable society.

Thus, with a clear vision and intelligent tools at our disposal, let us embark on this next chapter with a commitment to understanding, a thirst for knowledge, and a resolve to integrate the noblest of our aspirations into the everyday fabric of human exchange.

Ensuring Fairness and Inclusivity

In the pursuit of weaving social equity into the fabric of every transaction, fairness and inclusivity stand as our guiding stars. Ensuring that each exchange not only meets but exceeds the ideals of equity requires a meticulous approach, where the quality of contribution and merit of ideas are held in the highest regard. It's about crafting a system where the value provided is matched by the opportunities received, and where the scales of justice balance the inputs of effort with the outputs of reward.

Imagine Maya, a young aspiring programmer, struggling to find her footing in the competitive tech industry. Despite her talent and dedication, she faces an uphill battle against unconscious bias and a system that prioritizes experience over potential. Enter "EquitAble," a revolutionary AI-powered platform that utilizes a sophisticated algorithm to match employers with skilled individuals based purely on their **merit and ability**, not their resume or background. This platform becomes Maya's beacon of hope, connecting her with a thriving startup that values her unique perspective and potential, allowing her to showcase her skills and land her dream job.

Across the globe, in a bustling African village, a group of women farmers are struggling to access vital resources and fair market prices for their organic produce. A nonprofit organization, "Fair Trade Bridge," steps in, leveraging their network of ethical businesses and consumers to create a **fair trade marketplace**. This online platform empowers these women farmers by ensuring they receive a **fair price** for their crops, allowing them to invest in their families and communities, breaking the cycle of poverty and creating a more **equitable and prosperous** future.

These are just two examples of how the pursuit of social equity in transactions transcends mere statistics. It's about **human connections, dismantling barriers, and fostering opportunities**. Whether it's an AI-powered platform championing merit or a nonprofit organization creating fair trade avenues, the key lies in **harnessing technology** to amplify the power of **human values**.

In this quest for an equitable transactional landscape, we must leverage the best of our technologies and the breadth of our humanity. The analytical sharpness of AI, blended with the deep understanding of human values, can illuminate the path to fairer practices. These tools give us the power to scrutinize our systems, ensuring that merit is recognized and quality is rewarded, while also identifying and eliminating biases that hinder inclusivity.

Our transactions, whether they bridge continents or the space between two local businesses, must be reflective of a world where diversity is not just acknowledged but embraced as a source of strength and innovation. From the hiring practices of global corporations to the community initiatives that support local enterprises, the principles of fairness, quality, and merit must be paramount.

In a world where diversity is often seen as a challenge, we must recognize it as an **unparalleled strength**. From the global corporation that embraces diverse leadership teams to the local coffee shop that sources its beans from ethically sourced farms across the globe, the **principles of fairness**, **quality**, **and merit** must be the cornerstones of every transaction.

As we chart our course through the complexities of social transactions, our compass must be calibrated to the true north of fairness and inclusivity, guided by a vision that sees beyond the immediate to the potential of what can be achieved when every individual is given the chance to contribute and succeed based on their **true worth**. This is the synthesis of our efforts: to build a world where every transaction is a step toward a more just, prosperous, and equitable society.

Our journey toward social equity is a constant navigation, a continuous refinement of our systems and practices. We must remain guided by the compass of fairness and inclusivity, striving toward a future where every individual is valued for their unique contributions, where every transaction becomes a building block of a just, prosperous, and equitable society. In doing so, we unlock the true potential of every human being, creating a tapestry of diverse talents and perspectives that paves the way for a collective, brighter future.

Creating Shared Value

In the intricate dance of transactions that connect us, the creation of shared value is an essential step toward a harmonious future. This concept goes beyond the traditional metrics of profit and loss, expanding into the realms where quality and merit become the cornerstones of every interaction. It's about aligning our economic goals with the broader societal needs, forging a path where every handshake and every deal enriches our collective existence.

To build social equity into the very DNA of our transactions, we must draw from the wellspring of human ingenuity and the precision of our technological tools. The thoughtful integration of AI, with its ability to process and analyze vast troves of data, stands as a testament to our progress. Yet, it is the human touch – the philosophers' wisdom, the poets' empathy, the scientists' curiosity – that ensures these advancements serve the greater good.

In the bustling heart of Bangalore, India, a small team at "Eco-Weaves" are on a mission. They source handwoven textiles from rural artisans, ensuring fair wages and sustainable practices. Their intricately designed products are a testament to **quality and merit**, but their impact goes beyond aesthetics. Eco-Weaves partners with local NGOs to provide training and educational opportunities for the artisans, empowering them and strengthening their communities. This venture exemplifies the essence of **shared value**, where their commercial pursuit not only garners profit but also contributes to a **more equitable society**.

Across the globe, in a vibrant New York City coffee shop, Sarah, a barista brimming with passion, is crafting the perfect latte for a customer. While the transaction is seemingly simple, it embodies **shared value** in its own unique way. Sarah's **dedication and skill** are evident in the quality cup of coffee, but the interaction extends beyond caffeine. Sarah donates a portion of her tips to a local literacy program, demonstrating the human touch that complements technology and underlines the importance of social responsibility.

These stories illustrate the **intricate dance of transactions** woven with **quality, merit, and shared value**. We see this dance not just in grand ventures but also in everyday interactions. The key lies in **integrating human ingenuity with technological advancements**. AI algorithms can analyze data and trends, but it's the **philosophers' wisdom** that guides ethical decision-making, the **poets' empathy** that fosters inclusivity, and the **scientists' curiosity** that drives sustainable solutions.

Through the lens of shared value, we see that true wealth is not hoarded but distributed, not in scraps but in abundance that lifts all boats. The merit of an idea or the quality of a product must be weighed alongside its capacity to contribute to the public weal. Whether it's a startup pioneering sustainable practices or a conglomerate steering the global economy, the measure of their success is inextricably linked to the well-being they promote.

In this symphony of commerce and compassion, each of us plays a role. As thinkers and doers, creators and consumers, our transactions are the notes that can harmonize to create a melody of shared prosperity. We craft a narrative where every transaction is a building block for a world that values quality, rewards merit, and cherishes the collective advancement of society.

By drawing inspiration from these diverse voices, we can craft a narrative where every transaction is not just a business deal, but a **building block for a world that values quality, rewards merit, and cherishes collective progress**. This symphony of shared value, with each note contributing to the harmony, holds the potential to create a future where prosperity is not just measured in profits, but in the **collective well-being** of all.

What Are the Different Types of Transactions?

To weave social equity into the fabric of our transactions, we must first map the intricate web of institutions that underpin society, appreciating their distinct roles and the transactions they engage in. Consider the educational institutions – schools, colleges, and universities – each transaction here, whether it's a student's tuition fee or a community's investment in a new campus, is a seed planted for future intellectual harvests. Similarly, public services, from health organizations to law enforcement, carry out transactions that are the sinews and bones of our societal structure, each service rendered strengthening the collective body.

In the consumer realm, a transaction is not just an exchange of goods but a transfer of value that can ripple through the economy. A meal purchased at a restaurant, a subscription to a streaming service, or an investment in a startup by a venture capitalist – each of these is a stroke on the canvas of our economy, contributing to a vibrant picture of communal interdependence.

The SES, in this context, must be a dynamic, living score, reflective of the nuanced impact these transactions have. It should mirror the philanthropic investments of a Bill Gates, where funding for global health becomes a vector for widespread societal uplift, or the strategic insights of a Ray Dalio, where every financial move is a calculated step toward greater stability and equitable growth.

Incorporating AI and Generative AI, we could draw real-time examples like the AI-driven platforms that connect freelance workers with global opportunities, ensuring fair compensation and work opportunities regardless of geography – a modern echo of the author's vision for an inclusive technological ecosystem – or the GenAI systems that could optimize logistics for a transportation company, ensuring that a small business's products are delivered efficiently, reducing costs, and improving market competition.

Let's consider more tangible examples.

In the thriving heart of Mumbai, India, Maya, a single mother juggling multiple jobs, struggles to provide quality education for her son, Ravi. The local schools are overcrowded and underfunded, limiting Ravi's access to the resources he needs to thrive. Enter "Knowledge Weavers," an innovative social enterprise powered by AI. This platform utilizes a sophisticated algorithm to personalize learning experiences for students in underserved communities like Ravi's. By analyzing individual strengths and weaknesses, Knowledge Weavers curates a virtual classroom experience, connecting students with dedicated online tutors and interactive learning modules, all at an affordable cost. This initiative, reflecting the **democratization of access** advocated by AI and Generative AI, empowers Ravi and countless others like him, paving the way for a more **equitable education landscape**.

Across the Atlantic, in a small town in Vermont, Sarah, a passionate young farmer, strives to make her organic produce accessible to the local community. However, traditional distribution channels often leave small-scale farmers like Sarah with meager profits and limited market reach. Enter "Community Connect," a mobile app that utilizes AI algorithms to connect local consumers directly with ethical businesses like Sarah's farm. This platform not only empowers consumers to make informed choices about their food but also helps farmers like Sarah access a wider customer base and fair pricing for their products, fostering a more **sustainable and equitable food system**.

These real-time stories illustrate the intricate web of transactions that weave the fabric of our society, each carrying the potential to contribute to **social equity**. We see this not just in grand initiatives but also in everyday interactions. Just as **public services** like healthcare and law enforcement play a vital role in maintaining the **sinews and bones** of society, **consumer transactions**, from buying groceries to subscribing to streaming services, contribute to the vibrant tapestry of our **communal interdependence**.

The **Social Equity Score (SES)**, in this context, becomes a dynamic metric that reflects the **nuanced impact** of these diverse transactions. It mirrors the **philanthropic investments** of a Bill Gates, where funding for health initiatives in developing countries becomes a vector for **widespread societal uplift**. It echoes the **strategic insights** of a Ray Dalio, where responsible financial decisions pave the way for **equitable and sustainable growth**.

AI and Generative AI **play a crucial role in** scaling and amplifying **these efforts. AI-powered platforms connect freelance workers with global opportunities, ensuring** fair compensation **and** geographical inclusivity **in the workforce, echoing the author's vision. GenAI systems optimize logistics for small businesses, allowing them to compete effectively in the market.**

This narrative extends beyond the **financial realm**:

Regular services: An AI-enhanced education platform personalizes learning for students in remote regions, democratizing access to quality education.

Referral systems: A community-driven app that connects consumers with local businesses, amplifying the economic footprint of neighborhood enterprises.

Hybrid transactions: A media company offers bundled subscriptions, combining traditional news with modern entertainment, adapting to consumer preferences while preserving journalistic integrity.

Financial support: Microfinancing initiatives powered by blockchain, providing transparent, low-interest loans to women entrepreneurs in developing countries, embodying the thoughtful investment strategies praised by Ray Dalio.

Manual help: A volunteer matching system that uses AI to align skills with needs, placing a retired teacher in a mentoring role at a nonprofit, harnessing the ethos of service championed by Gates.

To distill this down into the SES, we envisage a system as multifaceted as the transactions it seeks to quantify. Like Lex Friedman's approach to AI, it would be discerning yet humanistic, ensuring that the "score" captures not only the economic but the social value, rewarding those who invest in people as much as in products. It would be a score that values the mentor's time as much as the financier's investment, recognizing that every transaction has a heartbeat, and every contribution, no matter how small, feeds into the larger narrative of our shared human experience.

Ultimately, building the SES is akin to composing a symphony in the style of the express thinkers – a harmonious blend of strategy, innovation, inclusivity, and intellect, where every note contributes to the grand crescendo of societal progress.

Monetary Transactions

In the realm of monetary transactions, the integration of social equity transcends ethical obligations, aligning profitability with a deeper sense of purpose. Envision a scenario where every financial exchange, from simple daily purchases to intricate corporate negotiations, carries an intrinsic awareness of its broader societal impact. Here, the Social Equity Score (SES), augmented by AI and Generative AI, emerges as a pivotal instrument.

By applying a principled approach, we can dissect monetary transactions for not just efficiency but fairness as well. For example, when evaluating a loan application, AI can be leveraged to ensure decisions are based on the merits, free from biases, thus refining the decision-making process.

Let's consider this:

In the vibrant streets of Nairobi, Kenya, James, a young entrepreneur, dreams of launching a sustainable clothing line, "Eco Threads." However, securing traditional bank loans proves difficult due to a lack of established credit history. Enter "Empower Capital," a revolutionary fintech company leading the charge in **socially responsible lending**. Their AI-powered platform analyzes not just financial data but also factors like community impact and environmental sustainability when evaluating loan applications. James's commitment to using recycled materials and empowering local artisans through fair wages resonates with Empower Capital's ethical framework, leading to the approval of a loan. This scenario exemplifies how AI can **refine decision-making**, ensuring **fairness and social equity** in financial transactions.

Across the globe, in a booming Sao Paulo market, Maria, a single mother juggling multiple jobs, struggles to make ends meet. Enter "MicroLoan Connect," a nonprofit organization utilizing AI-powered algorithms. This platform connects Maria with a network of microlenders, offering her access to **transparent and affordable** loan options. Additionally, through an integrated financial literacy program powered by Generative AI, Maria receives **tailored financial advice**, empowering her to manage her finances effectively. This story showcases how AI and Generative AI can be harnessed to **amplify financial equity** on a global scale, empowering individuals like Maria to break the cycle of poverty and achieve economic independence.

These stories illustrate the multidimensional nature of the **Social Equity Score (SES)**. It goes beyond simply measuring financial gain, encompassing the **enduring social impact** of every transaction. This concept encourages businesses to think beyond immediate profits and prioritize initiatives that contribute to **sustainable growth and societal well-being**.

With an optimistic view on technology, we see that such tools can amplify this equity on a global scale. Take a microloan issued to an entrepreneur in a developing region; by viewing the exchange through the SES framework, the lender can gauge its community impact, thereby investing in both social and financial returns. AI can forecast societal repercussions, modifying terms to support businesses with greater communal benefit.

A human-centric perspective on AI invites us to consider how these technologies can augment our human experience. In this context, AI might provide tailored financial advice to the borrower, promoting economic empowerment and ensuring the transaction meets ethical and fiscal standards.

The SES infuses transactions with multidimensional value, evaluating not just immediate financial results but also the enduring social advantages. It compels businesses to back initiatives that, while they may not yield the highest instant financial return, promise sustainable growth and societal well-being. A venture capitalist, for instance, could use SES metrics to favor an investment in an enterprise that offers affordable housing over a more profitable but less socially beneficial option.

Consider another scenario:

In the boardroom of a global energy company, a heated debate ensues. The company, facing pressure from shareholders, considers investing in a highly profitable but environmentally damaging oil exploration project. However, a young analyst, armed with SES data provided by AI, presents a compelling case for a seemingly less lucrative renewable energy project. This project, while offering lower immediate returns, boasts a significantly higher SES due to its positive environmental impact and potential for long-term societal benefit. Inspired by this data-driven approach, the board ultimately decides to invest in the renewable energy project, demonstrating how the SES can **reimagine monetary transactions** as drivers of **social innovation**.

Ultimately, the SES reimagines monetary transactions as conduits for social innovation. It demands a delicate equilibrium, balancing quality and merit against the prospective social dividends. By utilizing AI and Generative AI, we can ensure this balance is not only attained but also becomes the norm, propelling an economy that is as equitable as it is robust – an economy where technology is employed in the service of human advancement and equity.

Building an **equitable and robust** economy requires a delicate **equilibrium** between **quality, merit, and potential social dividends**. By leveraging AI and GenAI responsibly, we can ensure that this balance is not just attained but also becomes the norm. In doing so, we can weave a future where **technology becomes a powerful tool for human advancement and social equity**, ensuring that every transaction, big or small, contributes to a **more just and prosperous world for all**.

Nonmonetary Transactions

In the intricate tapestry of nonmonetary transactions, the threads of social equity are interwoven to form the fabric of communal solidarity. These transactions, though not financial, are potent catalysts for societal transformation. They encompass the generous exchange of knowledge, the voluntary transfer of skills, and the provision of support without the expectation of financial return.

Envision a platform where mentors freely offer guidance to the next generation of innovators, or where experts share valuable insights to empower communities. These exchanges, when seen through the lens of social equity, amplify their inherent value, fostering an ecosystem where merit and quality are the cornerstones of each interaction.

Let us quote some examples:

In the dynamic heart of Rio de Janeiro, Miguel dreams of becoming a renowned architect. However, access to quality architectural education is limited and expensive. Enter "Mentor Match," an innovative platform powered by AI. This platform connects aspiring individuals like Miguel with seasoned professionals based on **mutual interests and skill sets**. Miguel is matched with Ana, a renowned architect known for her commitment to sustainable design. Through **interactive online sessions and personalized guidance**, Ana mentors Miguel, fostering his talent and equipping him with the necessary skills to navigate the competitive world of architecture. This story illustrates how AI can **facilitate nonmonetary**

transactions that cultivate **merit and equip individuals with valuable knowledge** with the assumption that there are enough mentors to participate that requires significant effort in building awareness and encouraging mentors to participate without any monetary gain

Across the world, in a small village nestled in the Himalayas, a community struggles with limited access to clean water. Enter "Knowledge Share," a nonprofit organization utilizing AI-powered tools. This platform connects the village with a network of water engineers and sustainability experts across the globe. Through **interactive workshops and knowledge-sharing sessions**, the experts share their expertise on building sustainable water management systems. This transfer of **valuable skills and knowledge** empowers the community to address their needs and build a more **resilient future**, demonstrating how AI can contribute to **societal transformation**.

These stories showcase the crucial role of **nonmonetary transactions** in weaving the tapestry of social equity. These transactions are not merely exchanges; they are **catalysts for empowerment and transformation**. Obviously the organizations should be recognized and given incentives to encourage employees for any such societal contributions and participation. The **Social Equity Score (SES)**, when applied to this realm, becomes a **beacon for intentional community engagement**.

In the busy hub of a community center, a group of volunteers gathers to organize a food distribution drive. Traditionally, the volunteers relied on anecdotal evidence to assess the effectiveness of their efforts. However, this year, they utilize an AI-powered **impact assessment tool** developed by "Equity Track." This tool analyzes data from various sources, including the number of individuals served and the nutritional value of the distributed food, providing the volunteers with **measurable insights** into the **societal impact** of their efforts. This empowers them to **optimize their approach** and ensure that their **donations create the maximum possible impact**.

AI and Generative AI come into play as architects of a new landscape for these nonmonetary exchanges. They have the capacity to match mentors with mentees across the globe, optimizing the alignment of skills and needs. This technology can assess the quality of information shared in peer-to-peer learning networks, ensuring that the most accurate and beneficial knowledge is circulated.

AI can also facilitate the fair distribution of resources in volunteer networks, ensuring that the aid reaches those who need it most, and can measure the societal impact of such activities, enhancing the visibility of altruistic endeavors. Through Generative AI, we can simulate outcomes of various nonmonetary transactions, predicting their long-term benefits for individuals and communities alike.

By leveraging AI and Generative AI, we **transcend mere transactions** and delve into the realm of **transformational change**. These technologies allow us to **simulate potential outcomes**, predict the **long-term benefits** of nonmonetary transactions, and ensure that our **efforts are guided by the principles of equity, quality, and merit**. In doing so, we can foster a world where **knowledge and skills are shared freely**, where **resources are distributed fairly**, and where every **nonmonetary transaction becomes a building block for a more equitable and prosperous future for all**. We build a legacy of **shared value** that transcends the **transactional** and becomes **transformational**, ensuring that the **seeds of equity** sown today blossom into a **brighter tomorrow**.

The concept of social equity in this context becomes a beacon for intentional community engagement, encouraging a meritocracy of ideas and service. For instance, an AI-driven system could help optimize the distribution of donated resources, ensuring they create the maximum possible impact based on the needs of the recipients and the societal value of the donations optimizing the need vs availability.

What Are the Different Types of Service Weightage Efforts?

Imagine Sarah, a dedicated public health nurse, working tirelessly in a remote village in India. She not only provides critical medical care to the community but also spearheads educational workshops on hygiene and preventative healthcare. While Sarah's dedication is evident, quantifying the true impact of her work for the SES can be challenging. This is where AI steps in.

An AI-powered platform, called "Impact Mapper," utilizes real-time data and historical trends to analyze the **multidimensional impact** of Sarah's efforts. It not only considers the immediate health improvements in the community but also factors in the **long-term benefits** like increased school attendance due to improved health, and the potential for economic growth as the community becomes healthier and more productive. This comprehensive analysis allows for a **more nuanced understanding** of the **social value** Sarah's work generates, informing the weightage assigned to healthcare services within the SES.

Across the globe, in a bustling city like London, a vibrant community center thrives on the dedication of volunteers like Michael. He leads workshops on financial literacy and job search skills, empowering individuals from underprivileged backgrounds to break the cycle of poverty. However, it is difficult to accurately measure the impact of Michael's efforts solely through traditional methods.

Enter Generative AI. This technology allows for **simulating potential outcomes** of different service efforts. In Michael's case, Generative AI can analyze data on past participants who benefited from his workshops, simulating their potential employment prospects and increased earning potential. This data serves as a **powerful tool for assigning weightage** to community services like Michael's, highlighting the **indirect but significant contribution** they make toward social equity.

When we delve into the intricacies of assigning weightages to various service efforts in the context of a Social Equity Score (SES), we are essentially mapping out the blueprint of societal values and priorities. Each weightage assigned to a service category is not just a number; it's a reflection of the collective judgment about the importance of that service in promoting social equity. However, these weightages and the process of assigning them are susceptible to inherent biases and subjective interpretations, which is where AI and Generative AI come into play to enhance the realism and fairness in these assessments.

The impact of service efforts across different sectors – be it public health, education, or community services – varies significantly, and their contributions to social equity are complex and multidimensional. For instance, services in public health and safety are crucial for societal well-being and stability, hence their higher weightage. But how do we quantify the impact of a public health initiative compared to a community service program? This is where AI's analytical capabilities can provide deeper insights.

AI and Generative AI can dissect vast datasets to unearth the nuanced impacts of these services. By analyzing trends, outcomes, and correlations in historical and real-time data, AI can help us understand not just the direct impacts of services, but also their indirect, long-term contributions to society. For example, AI can evaluate how educational services influence economic mobility, or how community services enhance social cohesion, offering a more granular and dynamic perspective on their value.

Moreover, AI can identify and mitigate biases in the weightage assignment process. By learning from a broad spectrum of data sources and perspectives, AI can challenge human preconceptions and uncover overlooked aspects of services' contributions to equity. This objectivity and depth can help in fine-tuning the SES to more accurately reflect the true value of services to society.

AI and Generative AI also bring the advantage of adaptability. As societal values and priorities evolve, so too should the weightages in the SES. AI can continuously update these weightages based on new data and changing contexts, ensuring the SES remains relevant and reflective of current societal norms.

In leveraging AI and Generative AI to understand and adjust the service weightage efforts, we can achieve a more nuanced, objective, and realistic assessment of how various services contribute to social equity. This approach not only enhances the accuracy of the SES but also promotes a more informed and dynamic understanding of the interplay between different services and their collective impact on social equity.

By leveraging AI and Generative AI, the process of assigning weightages to service efforts within the SES becomes **more objective and multifaceted**. AI helps identify and **mitigate biases** that might skew the judgment of human assessors, while Generative AI provides valuable insights into the **long-term and indirect contributions** of different services. This ensures that the SES accurately reflects the true **social value** of diverse service efforts, fostering a more **informed and dynamic understanding** of their collective impact on **building a future where equity is the cornerstone of every interaction and every transaction**.

Individual Efforts

In the intricate tapestry of societal advancement, individual contributions through skill development and community service are pivotal. These efforts, when scrutinized and augmented by AI and Generative AI technologies, offer a dynamic lens to assess and magnify their influence on social equity. The interplay of individual growth and collective welfare is a critical arena where the power of AI can be harnessed to foster a more equitable society.

Skill development is not merely a pathway to personal advancement; it is a cornerstone for societal progress. It empowers individuals, equipping them with the tools to innovate, excel, and contribute meaningfully to their communities. This empowerment is doubly significant in the context of social equity, where the opportunity to develop and utilize skills can bridge economic disparities and catalyze upward social mobility.

Community service, in its essence, is the manifestation of individual commitment to societal health. It embodies the principle of giving back, reinforcing the social fabric through acts of kindness, support, and solidarity. These efforts, often grassroots and deeply personal, are fundamental to nurturing a sense of community and shared responsibility.

Imagine Amina, a young woman living in a remote village in Kenya. Despite limited resources, Amina is passionate about learning and driven to contribute to her community. However, traditional educational avenues are scarce in her village. Enter "Skills Spark," an AI-powered platform that utilizes a dynamic curriculum based on individual strengths and weaknesses. Through personalized learning plans and real-time skill gap identification, Amina can develop her computer literacy and coding skills, equipping her with the tools to compete in the global job market. This not only empowers Amina to pursue her personal aspirations but also allows her to contribute financially to her family and community, bridging the economic disparity often associated with limited educational opportunities.

Across the globe, in a vibrant city like New York, David, a retired engineer, feels a strong sense of responsibility to give back to his community. He volunteers at a local soup kitchen, offering his time and compassion to those in need. However, the impact of individual volunteers in large organizations can often feel intangible and difficult to measure.

This is where Generative AI steps in. An organization called "Impact Amplifier" utilizes this technology to **simulate the potential long-term benefits** of volunteer efforts. In David's case, Generative AI can analyze historical data on the soup kitchen's impact, including improved health outcomes and increased employment prospects for beneficiaries who received regular meals and support. This data provides a **quantifiable understanding** of David's individual contribution, highlighting the **ripple effect** of his compassion and dedication within the larger community.

By harnessing AI and Generative AI, the intricate interplay between **individual growth and collective well-being** becomes clearer and more impactful. These technologies don't just **analyze and assess**, they also **amplify and enhance** the power of individual efforts. In skill development, AI empowers individuals like Amina to reach their full potential, paving the way for greater economic mobility and social equity. In community service, AI empowers individuals like David to see the tangible impact of their actions, fostering a culture of **collective responsibility and shared progress**.

When AI and Generative AI are applied to these domains, they unlock unprecedented potential to analyze, predict, and enhance the impacts of individual actions on social equity. In skill development, AI can provide personalized learning pathways, identify skill gaps in real time, and offer predictive insights into future skill demands. This targeted approach ensures that skill acquisition is aligned with both individual aspirations and market needs, thereby enhancing the quality and relevance of personal and professional development.

For community service, AI can optimize resource allocation, match volunteers with opportunities that maximize their impact, and evaluate the effectiveness of various initiatives. By analyzing data on community needs and service outcomes, AI can help tailor community service efforts to be more responsive and impactful, ensuring that every act of service contributes optimally to community well-being.

Moreover, AI can assist in quantifying and recognizing the value of individual contributions in both domains. By establishing metrics that accurately reflect the quality and impact of personal efforts, AI can ensure that individuals are acknowledged and rewarded for their contributions, fostering a culture of meritocracy and fairness.

In this nuanced dialogue between individual efforts and societal benefits, AI and Generative AI emerge as crucial allies. They offer tools not just for analysis but for enhancement, ensuring that individual pursuits in skill development and community service are not just recognized but also optimized for the greater good. This synergy between personal endeavor and technological innovation paves the way for a future where social equity is actively cultivated and celebrated, rooted in the quality and merit of individual contributions.

Ultimately, AI and Generative AI are not replacements for human ingenuity and compassion, but rather **powerful allies** in building a more equitable society. They act as facilitators, ensuring that **individual talent and dedication** are recognized, nurtured, and translated into meaningful contributions, weaving a tapestry of **social equity** where the quality and merit of every individual effort resonate throughout society.

Organizational Efforts

In the landscape of modern corporate ethos, the integration of corporate social responsibility (CSR) and employee volunteer programs stands as a beacon of organizational commitment to social equity. These initiatives, when thoughtfully executed, not only bolster the corporation's image but also instigate profound ripple effects on societal well-being. Leveraging AI and Generative AI, organizations can elevate these programs from mere acts of goodwill to strategic endeavors that significantly enhance social equity while aligning with corporate objectives.

Corporate social responsibility, in its evolved form, transcends the traditional boundaries of philanthropy. It represents a holistic commitment by organizations to contribute positively to society, encompassing environmental stewardship, ethical business practices, and economic development. In this realm, AI can serve as a potent tool to analyze societal needs, tailor CSR initiatives to address those needs effectively, and measure the impact of these initiatives with precision.

By doing so, corporations can ensure that their CSR efforts are not only generous but also strategically aligned with the areas of greatest need, thereby maximizing their contribution to social equity.

Employee volunteer programs represent another critical dimension of organizational efforts toward social equity. These programs, when effectively harnessed, can transform the collective energy and skills of the workforce into a powerful force for social good. AI and Generative AI can revolutionize these programs by identifying optimal matching opportunities for employee skills and community needs, thereby enhancing the quality and impact of volunteer efforts. Furthermore, AI can track and analyze the outcomes of these volunteering initiatives, providing valuable feedback and insights to refine and intensify future efforts.

The use of AI in these contexts also introduces a merit-based framework to assess and recognize contributions toward social equity. By leveraging data analytics and AI algorithms, organizations can quantify the impact of their CSR and volunteer initiatives, attributing value to these efforts based on their outcomes rather than just their inputs. This approach not only fosters a culture of accountability and effectiveness but also motivates continuous improvement and innovation in social responsibility endeavors.

Moreover, AI can assist in aligning these organizational efforts with broader societal goals, such as the United Nations Sustainable Development Goals (SDGs). By analyzing vast datasets and generating insights into global and local challenges, AI can guide organizations in directing their CSR and volunteer efforts toward areas where they can make the most significant difference, thereby enhancing the quality and relevance of their contributions to social equity.

Let's quote a couple of examples to get the context a bit more:

In the energetic heart of Bangalore, India, "GreenTech Solutions," a leading renewable energy company, grapples with fulfilling its corporate social responsibility (CSR) goals. While the company champions sustainability, achieving tangible impact in underprivileged communities can be challenging. Enter "Impact Compass," an AI-powered platform that utilizes **real-time data analysis** to identify the most pressing needs in neighboring villages.

Through Impact Compass, GreenTech discovers that lack of access to clean water is a major concern in a nearby village. This insight allows them to tailor their CSR initiative, leveraging their expertise in solar power to install solar-powered water purification systems. This initiative not only addresses a crucial need in the community but also aligns with GreenTech's core values of sustainability and innovation. This exemplifies how AI can help organizations **strategically align CSR efforts** for **maximum societal impact**.

Across the globe, in a vibrant city like San Francisco, "Technovation" is a software company committed to fostering diversity and inclusion within its workforce. However, ensuring their employee volunteer program effectively utilizes the diverse skill sets of their employees can be complex. This is where Generative AI steps in.

"SkillMatch," an AI-powered platform developed by Technovation, analyzes employee skill sets and preferences, matching them with volunteer opportunities that best suit their expertise and interests. This personalized approach ensures that employees contribute their skills effectively, maximizing the **quality and impact** of their volunteer efforts. Additionally, SkillMatch tracks participation data, generating **feedback and insights** that help Technovation refine their volunteer program and celebrate the **meritorious contributions** of their employees.

By leveraging AI and Generative AI, organizations like GreenTech and Technovation are **elevating their social responsibility initiatives from acts of goodwill to strategic endeavors** that **significantly enhance social equity**. AI empowers organizations to **analyze needs, tailor programs, and measure impact** with precision, ensuring that their efforts are **not just generous but also impactful and aligned** with both corporate objectives and the broader societal goals like the UN SDGs. This **merit-based framework** fosters **accountability, effectiveness, and continuous**

improvement, paving the way for a future where **corporations become powerful change-makers**, contributing to a more **equitable and sustainable world**.

In conclusion, the strategic integration of AI and Generative AI in corporate social responsibility and employee volunteer programs offers a transformative pathway for organizations to amplify their impact on social equity. By leveraging these technologies, organizations can ensure that their efforts are not only well-intentioned but also strategically aligned, impactful, and meritorious, thereby contributing to a more equitable and sustainable future.

How Do You Review and Rate a Transaction?

In the pursuit of developing a nuanced and equitable Social Equity Score (SES), it's imperative to consider a multifaceted approach that incorporates various factors and variables from individuals, entities, and organizations involved in transactions. This meticulous approach ensures the removal of bias, fostering a fair and balanced assessment grounded in quality and merit, with AI and GenAI serving as critical tools in this evaluative process.

For transactions, a tiered system categorizes them into referral, regular services, hybrid models, financial support, and manual help, each with its base score and potential for additional points based on percentage ratings.

This tiered structure acknowledges the inherent value and impact of different transaction types:

Referral transactions are foundational, yet their value can significantly increase based on the effectiveness and outcomes of the referrals, meriting an initial score with potential for enhancement based on performance ratings.

Regular services form the backbone of service transactions, with their score reflecting the immediate value provided and an additional component that rewards service excellence and customer satisfaction.

Hybrid transactions, which blend various service modalities, receive a higher base score reflecting their complexity and value-add, with room for upward adjustment based on effectiveness and client feedback.

Financial support transactions are pivotal in their ability to drive growth and stability, deserving a higher base score, with distinctions made between investments, microfinancing, and loans to reflect their varying levels of risk and impact.

Manual help represents a direct, often personal contribution to welfare, meriting the highest base score, recognizing the human effort and empathy involved.

The SES calculation (SEU = Social Equity Utility) incorporates these base scores and introduces multipliers based on service effort ratings, individual status, and disability or senior status. This detailed formula ensures that each transaction's score is reflective not just of its category but of its quality, execution, and the personal circumstances of those involved.

In this equation, variables such as age, economic status, and experiential factors like being a victim of a calamity are considered, adding layers of context that ensure the SES is as nuanced and representative as possible. The ratings within each category allow for a dynamic scoring system that adjusts to reflect both the transaction's inherent value and its execution quality.

Now Imagine Maria, a single mother of two in a bustling city like Rio de Janeiro, struggling to make ends meet. She works tirelessly as a domestic helper, juggling multiple jobs to support her family. Enter "Helping Hands," a nonprofit organization utilizing the Social Equity Score (SES) to connect individuals like Maria with various support services.

Helping Hands leverages AI to analyze Maria's situation, considering her economic status, single parent status, and demanding work schedule. Based on this information, Maria is categorized as a "vulnerable individual" within the individual status variable, which adds a multiplier to her SES score when she engages in transactions with organizations like Helping Hands.

Maria approaches Helping Hands seeking financial support and childcare services. The "financial support" category carries a higher base score due to its potential to drive growth and stability for individuals like Maria. Additionally, as a single parent experiencing economic hardship, Maria receives additional points due to her specific circumstances.

However, Helping Hands doesn't solely rely on the base score. AI analyzes data on past participants who received microloans and childcare support, allowing Generative AI to **simulate the potential long-term impact** on Maria's financial stability and her children's well-being. This simulation, combined with Maria's positive service ratings after receiving support, further contributes to her final SES score in this transaction.

The "childcare services" transaction falls under the "regular services" category. While it carries a lower base score compared to financial support, the quality of care and Maria's satisfaction with the service also contribute to the final score. AI helps Helping Hands track these factors, ensuring both the base score and additional points reflect the true value and impact of the service provided.

Thanks to the financial support and reliable childcare services, Maria was able to pursue further training and education, leading to a better-paying job. Over time, her SES improved, reflecting not just the immediate help she received but the long-term positive changes in her life.

This example showcases how the SES goes beyond a simple numerical evaluation. By considering a multifaceted approach that incorporates individual circumstances, service quality, and potential long-term impact, the SES, powered by AI and Generative AI, strives to create a **fair and balanced system**. This fosters a more **equitable ecosystem** where individuals like Maria receive the support they deserve, regardless of their starting point, and where every transaction contributes to a society grounded in **quality, merit, and the value created for all involved.**

AI and GenAI technologies come into play by analyzing vast datasets to identify patterns, trends, and outliers in these transactions, ensuring that the SES is not only accurate but also adaptive to real-world complexities and changes. These technologies can help refine the weighting of different variables, predict the long-term impacts of transactions, and continually update the scoring system to remain relevant and fair. This holistic approach ensures that the support provided leads to meaningful and sustainable improvements in the lives of individuals like Maria.

In sum, this approach to building an SES is about creating a system that is comprehensive, dynamic, and fair, leveraging advanced technologies to ensure that every score is a true reflection of value created or exchanged, thereby fostering a more equitable and meritorious ecosystem.

Establishing Criteria for Evaluation

Refining the evaluation criteria for the Social Equity Score (SES) involves a thoughtful integration of diverse factors including economic status, experiences of calamity, age, gender, and status, alongside the quality and merit of individual and collective experiences. This enriched assessment framework ensures a comprehensive and just evaluation, meticulously considering each element to capture the multifaceted nature of social equity. By embedding social impact into the evaluation's core, the SES transcends numerical assessment, offering a nuanced reflection of true equity and fairness across various dimensions of life and society.

In this detailed framework, age becomes a factor not as a bias but as a context, recognizing the varying needs and contributions of different life stages. Gender is considered to ensure inclusivity and to address any disparities, making the SES a tool for promoting gender equity. Marital or social status further refines the score, acknowledging the diverse roles and responsibilities individuals may have, which could influence their ability to contribute to or benefit from social equity initiatives.

Economic status is a critical factor, acknowledging that financial resources and constraints significantly influence one's ability to engage with and contribute to society. By factoring in economic status, the SES can more accurately reflect individuals' contexts, offering insights into the economic barriers or advantages that impact their social contributions and experiences.

The experience of being a victim of a calamity – be it natural disasters, socioeconomic crises, or personal tragedies – adds another layer to the SES. This consideration ensures that individuals and communities affected by adverse events are recognized for their resilience and are supported in ways that acknowledge their specific circumstances and needs.

Quality is another pivotal aspect, extending to the quality of experiences individuals have in their professional, personal, and societal interactions. For professionals, this might mean evaluating the impact of their work environment, career development opportunities, and professional relationships. Personally, the quality of one's relationships, education, and health can significantly influence their social equity standing. Societally, the quality of community engagement, civic participation, and access to public services are essential determinants.

Incorporating these factors requires a nuanced approach, where AI and GenAI can play a transformative role. These technologies can analyze complex datasets to assess economic status, identify individuals affected by calamities, and evaluate the quality of professional, personal, and societal experiences. They can discern patterns and correlations that might not be immediately apparent, providing a more comprehensive and detailed assessment.

In the vibrant city of Lagos, Nigeria, Chioma, a young entrepreneur, navigates the busy marketplace. While passionate about her business selling handmade jewelry, she faces challenges accessing resources and training opportunities due to her **economic status** as a **single mother**. Enter "EquityNet," an organization using an **enhanced Social Equity Score (SES)** that goes beyond mere numbers.

EquityNet utilizes AI to analyze Chioma's situation. Her **age** is considered as context, recognizing her potential at this stage in her career. Her **gender, marital status**, and **economic status** are also factored in, acknowledging the unique challenges she faces as a single entrepreneur. Considering her limited access to resources, EquityNet recognizes Chioma's **resilience** and **determination**, factors previously absent in the SES, thanks to the inclusion of **experiences of calamity**.

Chioma's dedication to her craft is evident in the high quality of her handmade jewelry. This translates into positive reviews from customers, contributing to the "quality" aspect of her SES score. However, the SES also factors in the **quality of her professional environment**. EquityNet assists Chioma in connecting with other female entrepreneurs, fostering a supportive network and improving her overall professional experience.

This enhanced SES, powered by AI and Generative AI, allows EquityNet to provide Chioma with **actionable insights**. By analyzing data from similar female entrepreneurs and those who overcame similar challenges, Generative AI can **simulate the potential impact** of various support services. Based on these simulations and Chioma's specific needs, EquityNet connects her with access to microloans, business training workshops, and mentorship opportunities.

Across the globe, in a bustling city like London, David, a seasoned IT professional, contemplates his career path. While successful, he feels a growing unease with the limited diversity in his workplace. This lack of **inclusivity** negatively impacts the **quality of his societal experience**.

David decides to volunteer his expertise at "Bridge Builders," a nonprofit organization promoting diversity and inclusion in the workplace. His contributions are evaluated using the enhanced SES, which now considers the **quality of societal engagement** through factors like promoting inclusivity. David's dedication to mentoring underrepresented individuals and fostering a diverse work environment significantly contributes to his SES score.

Through AI, Bridge Builders can analyze data on companies with inclusive work environments and the long-term benefits for both the company and its employees. This allows them to **recommend best practices** to David and other volunteers seeking to improve societal equity and inclusion within their organizations.

By incorporating diverse factors beyond just numerics, the enhanced SES, powered by AI and Generative AI, offers a more **holistic and responsive evaluation**. This leads to **actionable insights** and **targeted support**, empowering individuals like Chioma to overcome challenges and thrive, while inspiring individuals like David to contribute to a more **equitable and inclusive society**, ultimately building a future where **true social equity** goes beyond a score, but becomes a vibrant reality in every interaction and experience.

Moreover, AI and GenAI can adapt the SES criteria to reflect professional, personal, and societal norms, ensuring that the score remains relevant and aligned with current standards and expectations. By continuously updating the criteria based on real-world data and evolving norms, these technologies ensure that the SES is a dynamic and accurate reflection of social equity.

In sum, by integrating these additional concepts into the SES evaluation, supported by the analytical power of AI and GenAI, we can construct a more holistic and responsive measure of social equity. This measure not only reflects the diverse experiences and circumstances of individuals and communities but also provides actionable insights to improve equity across professional, personal, and societal spheres.

Gathering Feedback and Data

In the intricate process of calculating the Social Equity Score (SES), gathering nuanced feedback and comprehensive data is crucial. This involves deploying additional surveys and interviews to capture a wide array of perspectives and experiences, ensuring that the SES is grounded in the real-world contexts of those it aims to represent. Beyond traditional data collection, further monitoring and evaluation mechanisms are essential to validate and refine the SES, maintaining a steadfast focus on quality and merit.

Integrating AI and Generative AI into this process transforms the depth and breadth of analysis possible. These technologies can sift through extensive datasets derived from surveys and interviews, extracting key insights that might elude manual analysis. They enable a dynamic approach to monitoring and evaluation, where data is not just collected but continually interpreted in real-time, allowing for ongoing adjustments to the SES framework.

Moreover, AI's predictive capabilities can forecast the potential impacts of various social equity initiatives, guiding organizations in strategizing more effectively. By simulating different scenarios, Generative AI can help in understanding the long-term implications of specific SES scores, offering a foresight that aids in strategic planning and policy formulation.

Let us take an example here: In the thriving heart of Mumbai, India, the "Empowerment Network" faces a crucial challenge. They are dedicated to promoting social equity, but their current methods of gathering data for the SES rely heavily on traditional surveys, which can be time-consuming and potentially miss valuable insights. Enter "Voice of Change," a cutting-edge platform powered by AI and Generative AI.

Voice of Change utilizes sophisticated AI algorithms to analyze social media data in real-time. This captures a broader spectrum of perspectives and experiences compared to traditional surveys, allowing the Empowerment Network to better understand the needs and challenges faced by various communities. For instance, AI can analyze the sentiment of online discussions about access to education or healthcare, revealing areas where disparities might exist and informing targeted initiatives.

Beyond social media, Voice of Change also facilitates interactive interviews. By utilizing conversational AI, the platform conducts interviews in a more personalized and engaging manner, encouraging participants to share their experiences in greater depth. This approach allows for capturing the nuances of individual stories and the emotional impact of social inequities, which traditional surveys often lack.

However, data collection is just one aspect of the process. Voice of Change also employs Generative AI to predict the potential impact of various social equity initiatives. By simulating different scenarios based on real-world data, the platform can forecast how specific interventions might affect SES scores in different communities. This forward-looking approach allows the Empowerment Network to strategically allocate resources and prioritize initiatives with the highest potential for positive change.

For example, the Empowerment Network might be considering a community-based education program aimed at improving literacy rates. By leveraging Generative AI, they can simulate the potential impact of this program on the SES scores of children in different age groups and socioeconomic backgrounds. This data-driven approach helps the Network make informed decisions about resource allocation and program design, maximizing their impact on social equity.

The integration of AI and Generative AI into the SES data collection and evaluation process not only enhances its depth and breadth but also ensures a focus on quality and merit. These technologies offer a valuable set of tools for gathering comprehensive and objective data, while predicting the potential impact of interventions. Ultimately, this empowers organizations like the Empowerment Network to create a more insightful and actionable SES, one that serves as a powerful tool in their pursuit of a fairer and more equitable world.

In this sophisticated data ecosystem, the emphasis on quality and merit remains paramount. AI and GenAI serve as invaluable allies in ensuring that the SES remains a reliable and robust measure, one that is sensitive to the nuances of individual and collective experiences while upholding the highest standards of accuracy and objectivity.

Through this technologically enhanced approach, the process of gathering feedback and data for the SES becomes not just a methodical exercise but a strategic endeavor. It harnesses the power of advanced analytics to forge a score that is truly reflective of social equity, providing organizations and societies with a tool that is both insightful and actionable in their quest to foster a fairer world.

Implementing Improvements

Implementing improvements in the Social Equity Score (SES) calculation is a dynamic process that necessitates addressing identified issues and fostering an environment of continuous growth and learning. This iterative approach ensures that the SES remains a relevant and effective tool for measuring and enhancing social equity. By pinpointing areas that require enhancement and integrating feedback loops, the SES can evolve to more accurately reflect the complexities of social equity, always anchored in the principles of quality and merit.

The integration of AI and Generative AI plays a pivotal role in this refinement process. These technologies offer sophisticated data analysis capabilities, identifying patterns and anomalies that may not be apparent to human evaluators. AI can highlight areas within the SES framework that need adjustment, suggest modifications, and predict the outcomes of potential changes, thereby informing a strategic approach to improvements.

Furthermore, AI and GenAI enable a proactive stance on learning and adaptation within the SES calculation process. They facilitate the assimilation of new information and insights, ensuring that the SES remains adaptable and forward-looking. This adaptability is crucial for the SES to stay aligned with evolving societal norms and expectations, ensuring its continued relevance and utility.

In this context, the focus on quality and merit is paramount. The enhancements implemented must not only address identified gaps or issues but also elevate the SES's overall integrity and value. AI and GenAI can ensure that these enhancements are grounded in empirical evidence and aligned with the core objectives of social equity, maintaining the SES's commitment to fairness and excellence.

By adopting this approach, the process of updating and improving the SES becomes a model of continuous innovation and precision. It exemplifies a commitment to upholding the highest standards of social equity measurement, leveraging cutting-edge technology to ensure that the SES is a robust, insightful, and impactful tool in the quest for a more equitable society.

Let's take an example: Imagine a busy community center in Chicago called "Bridgeway," dedicated to fostering social equity and inclusivity. Their mission is noble, but their current Social Equity Score (SES) calculation relies on static data and doesn't fully capture the nuances of individual and community experiences. Enter "Empowerment Labs," a team of AI specialists pioneering advancements in the SES framework.

Empowerment Labs collaborates with Bridgeway to identify areas for improvement. Through **sophisticated AI algorithms**, they analyze data from various sources, including surveys, interviews, and even social media sentiment analysis. This **multidimensional approach** allows them to identify areas where the current SES might be underestimating or overlooking certain aspects of social equity.

For example, AI might reveal a disparity in access to quality healthcare within a specific demographic, despite the community center offering health education workshops. This insight prompts Bridgeway to explore the reasons behind the disparity, which could lead to targeted initiatives like partnering with local health providers to offer subsidized checkups or transportation assistance.

Beyond identifying gaps, AI plays a crucial role in **predicting the potential impact** of proposed changes. By utilizing **Generative AI**, Empowerment Labs can **simulate various scenarios** based on real-world data. This allows Bridgeway to explore the potential benefits and unintended consequences of different interventions before implementing them.

For instance, Bridgeway might consider launching a new after-school program focused on STEM education for underprivileged youth. Generative AI can simulate the program's potential impact on the SES scores of participating children, factoring in factors like improved academic performance, increased access to future career opportunities, and potential social mobility.

This **data-driven approach** empowers Bridgeway to make **informed decisions** and **prioritize initiatives** with the highest potential for positive change. Additionally, AI and Generative AI facilitate a **continuous feedback loop**. As Bridgeway implements new initiatives and gathers new data, these technologies can analyze the resulting impact and suggest further refinements to the SES framework.

This **iterative process**, combined with a relentless focus on **quality and merit**, ensures that the SES remains a **relevant and effective tool** for Bridgeway and other organizations striving for social equity. AI and Generative AI empower them to **continuously learn, adapt, and refine** their approach, ultimately contributing to a **more equitable and inclusive society** where individual experiences and contributions are recognized and valued.

Conclusion

In concluding this chapter, we've navigated the nuanced terrain of integrating metrics and measurements into the fabric of social equity. The journey through establishing, refining, and implementing the Social

Equity Score (SES) underscores a commitment to a data-driven, methodical approach in quantifying and enhancing social equity. At its core, the SES serves as a compass, guiding entities and individuals in understanding and amplifying their contributions to a more equitable society. The calibration of this score, rooted in the principles of quality and merit, reflects a harmonious blend of rigorous analytics and a deep commitment to fairness.

The SES calculation, distilled to its essence, is an intricate alchemy of diverse variables and factors, each meticulously weighed and interwoven. This high-level process is underpinned by advanced technologies – AI and GenAI – that offer unparalleled insights and foresight, ensuring that the SES is both reflective of current realities and adaptive to future shifts. These technologies, in their role as analytical and predictive aides, ensure that the pursuit of social equity is grounded in objectivity, precision, and a forward-looking ethos.

As we transition to the next chapter, our focus shifts to unraveling the "secret sauce" behind the SES. We will delve into the mechanics of how the SES operates within the intricate web of transactions among individuals, entities, and organizations. This exploration will illuminate the sophisticated interplay of variables that the SES captures, offering a deeper understanding of its construction and application. We will dissect the methodologies, the algorithms, and the analytical frameworks that constitute the backbone of the SES, providing clarity on how this powerful tool distills complex interactions into a coherent, actionable score.

In this forthcoming chapter, we will peel back the layers of the SES, revealing the intricate workings of its mechanics. By demystifying the processes that underpin the SES, we aim to equip readers with a robust understanding of how this score can be leveraged to foster, measure, and accelerate the journey toward genuine social equity. Through this exploration, we will continue our commitment to blending rigorous analysis with a deep-seated dedication to equity and justice, charting a course toward a future where every transaction is imbued with the potential to contribute to a fairer and more equitable world.

CHAPTER 7

Secret Sauce Using AI

Diving into the realm of Social Equity Score (SES) construction, this chapter unravels the sophisticated "secret sauce" that leverages AI and GenAI to craft a nuanced, dynamic scoring system. The foundational concepts, variables, and frameworks laid out in the previous chapter set the stage for a deep dive into the advanced methodologies that AI brings to the table in refining and applying these scores.

Artificial intelligence (AI) and its advanced counterpart, Generative AI (GenAI), are at the forefront of transforming abstract concepts and raw data into actionable insights. These technologies are adept at sifting through complex layers of information, identifying patterns, and predicting outcomes with a precision that is indispensable in the context of social equity.

The "secret sauce" lies in the AI-driven process that intricately analyzes various dimensions of social interactions, transactions, and contributions to calculate the SES. This process involves a detailed examination of the quality and merit of actions, ensuring that the score is not just a reflection of quantity but a testament to the impactful and meaningful contributions toward equity.

AI algorithms are tailored to assess a wide array of data points – from individual achievements and organizational efforts to broader societal impacts – ensuring a comprehensive evaluation. GenAI goes a step further by simulating potential future scenarios, providing a predictive perspective on how different actions and strategies might shape the SES.

© Raghu Banda 2024
R. Banda, *Building Social Equity with AI*, https://doi.org/10.1007/979-8-8688-0091-7_7

This chapter will elucidate how AI and GenAI synthesize vast datasets, apply sophisticated analytical models, and generate scores that are reflective of genuine contributions to social equity. It will detail the mechanisms through which these technologies account for nuances, mitigate biases, and uphold the principles of fairness and accuracy.

Moreover, the discussion will extend to how AI-infused processes can forge lasting connections and networks, underpinned by a shared commitment to social equity. By continuously learning and adapting, AI and GenAI ensure that the SES remains relevant, responsive, and robust over time, providing a reliable metric for individuals and organizations to gauge and guide their efforts in fostering a more equitable society.

In essence, this chapter aims to demystify the advanced AI methodologies that convert theoretical frameworks and raw data into a coherent, dynamic SES. It will highlight how the integration of AI and GenAI in this process not only elevates the precision and relevance of the score but also embeds a culture of continuous improvement and strategic foresight in the pursuit of social equity.

Let's take an example in how we deal with this: In the bustling heart of Rio de Janeiro, "Caminhos Cruzados" (Crossed Paths) is a nonprofit dedicated to promoting social equity. While their current methods for calculating the SES rely on static data, it often overlooks the intricate tapestry of individual experiences. Enter "Empowering Insights," a revolutionary platform powered by AI and GenAI.

Empowering Insights utilizes AI to analyze data from various sources, including employment records, community outreach programs, and even social media engagement. By sifting through this **multilayered information**, AI can identify patterns and nuances that might be missed by traditional methods.

For example, AI might uncover that a specific community center, despite offering educational workshops, faces a disparity in literacy rates. This insight prompts Caminhos Cruzados to explore the **underlying**

reasons, such as lack of transportation or childcare options. Armed with this knowledge, they can work with local partners to address these barriers, creating a more **equitable access to education**.

Beyond analysis, **GenAI** offers the power of **simulating potential future scenarios**. Imagine Caminhos Cruzados considering a new after-school program focused on financial literacy for young adults. GenAI can **predict the program's impact** on the SES, factoring in based on factors like improved financial decision-making, increased economic opportunities, and potential ripple effects throughout the community.

This **data-driven foresight** empowers Caminhos Cruzados to make **informed decisions** and **prioritize initiatives** with the highest potential for positive change. Additionally, AI and GenAI facilitate **continuous learning and adaptation**. As Caminhos Cruzados implements new programs and gathers new data, these technologies can analyze the resulting impact and suggest further refinements to the SES framework, ensuring its ongoing **relevance and effectiveness**.

This **dynamic approach**, fueled by AI and GenAI, fosters a **culture of continuous improvement**, ensuring Caminhos Cruzados remains a **reliable and robust tool** in the ongoing pursuit of social equity. As individuals and organizations strive to build a more equitable society, this chapter unveils the "secret sauce" – the transformative power of AI and GenAI, not just in calculating scores, but in **shaping a future woven with understanding, foresight, and lasting impact**.

What Is the Secret Sauce (with Crypto) in Building the Score?

In this exploration of the "secret sauce" behind AI-driven Social Equity Scores (SES), we delve into the sophisticated algorithms and processes that leverage blockchain technology to ensure transparency, security, and immutability in SES calculations. This chapter will dissect how these advanced technologies are employed to build, refine, and implement a dynamic SES that evolves over time, reflecting the genuine social impact of individuals and organizations.

At the heart of the SES is a nuanced consideration of various factors that influence social equity – ranging from environmental contributions and social responsibilities to ethical practices. Blockchain's role is pivotal, providing a decentralized ledger that meticulously records every transaction and interaction, forming the backbone of the SES. This ledger ensures that every data point contributing to the SES is verifiable and immutable, enhancing the trustworthiness and reliability of the score.

AI-driven algorithms, built upon this robust blockchain foundation, are designed to sift through complex datasets, identifying key patterns, trends, and insights that are critical for SES calculations. These algorithms can dissect massive volumes of data to pinpoint areas where individuals and organizations can improve their social impact, offering tailored suggestions that are grounded in empirical evidence.

Moreover, these AI systems provide dynamic recommendations for enhancing social equity scores by analyzing past performances and predicting future trends. They consider a broad spectrum of data, including feedback from additional surveys and interviews, continuous monitoring, and evaluations, ensuring that the SES remains a current and accurate reflection of social equity contributions.

The SES also benefits from AI's ability to facilitate lifelong connections by spotlighting opportunities for individuals and organizations to amplify their social impact. This not only aids in building a robust reputation but also fosters stronger community ties and stakeholder relationships, contributing to a sustainable and positive societal footprint.

Moving forward, we will delve deeper into the mechanics behind the SES. We'll unpack the intricate details for data integrity and utilize sophisticated algorithms to calculate and enhance the SES. This discussion will illuminate the processes that ensure the SES is a fair, objective, and actionable tool, grounded in the principles of quality and merit, and how it can be strategically utilized to foster and measure genuine social equity.

The following process outlines the steps involved at a very high level and leaves it to the reader's implementation of how SES can be adapted. In the last chapter, we shall provide references to the ***author's upcoming startup*** that is focused on building this SES.

The algorithm for calculating the Social Equity Score (SES) is a multistep process that leverages a combination of input and calculated variables. Here is an expanded explanation of each step based on the given pseudocode and steps: (**Demystifying the Social Equity Score (SES) Calculation: A Step-by-Step Breakdown**).

The provided pseudocode outlines the calculation process for a Social Equity Score (SES), aiming to assess the social contributions of individuals or organizations.

Here's a breakdown of the steps involved, simplified for easier understanding:

1. **Setting the stage:**

    - The algorithm requires several **input variables** such as the number of services provided (SP), received (SR), and their categories (SC).

    - Additional factors like service efforts (SE) and their associated ratings are also considered.

2. **Calculating service factors:**

    - The **service factor (Sfac)** is the initial indicator of contribution, calculated by dividing **services provided** by **services received**. Higher Sfac indicates more services provided.

Service factor (SFac) calculation: This is the ratio of services provided (SP) to services received (SR). It provides a preliminary assessment of an entity's engagement level within a network.

3. **Assigning weights and effort values:**

   - Each **service category (SC)** is assigned a **weightage (SW)** based on its perceived social impact. For example, "Charity & Community Services" gets a higher weightage than "Retail centers."

   - **Service efforts (SE)** are assigned different values based on their type (referral, regular, etc.) and adjusted further based on the service rating (1–100). Additionally, individual and disability/senior status ratings are factored in, reflecting individual circumstances. This combined value is called **service efforts updated (SEU).**

   Service category weightage (SW) calculation: Each service category (SC) is assigned a numerical value, which is then multiplied by a predetermined weight (150 in this example), to reflect the importance or impact of that category.

4. **Balancing the score:**

   - The **average of service weightage and service efforts updated (Avg_SW_SEU)** is calculated, essentially averaging the importance and extent of services provided and received.

   - To ensure the score falls within a specific range (150–750), a **minimum maximum number (MinMax_Num)** is calculated by subtracting 150 from the Avg_SW_SEU.

Average service weightage and efforts calculation (Avg_SW_SEU): The average of the weightage (SW) and the updated service efforts (SEU) is computed to obtain a nuanced view of the service's value, considering both its category and the efforts put into it.

Min Max number calculation (MinMax_Num): This step ensures that the SES remains within an acceptable range (150 to 750). The MinMax_Num is derived by subtracting a base number (150) from Avg_SW_SEU, effectively scaling the average to fit within the desired range.

5. **Adjusting for service imbalance:**

   - The **Avg of Min Max number and SFac (Avg_ MM_Fac)** is calculated differently based on the **Sfac** value:

     - If **more services are provided (Sfac > 1):**

       - If the adjusted MinMax_Num is less than 200, use 200 as Avg_MM_Fac for consistency.

       - Otherwise, use the adjusted MinMax_Num itself.

     - If more services are received (Sfac < 1):

       - Multiply the adjusted MinMax_Num by Sfac to proportionally adjust the score.

   Average Min Max factor (Avg_MM_Fac) calculation: The Avg_MM_Fac is adjusted based on whether the entity has provided more services than received (SFac > 1) or received more than provided (SFac < 1). For SFac > 1, the Avg_MM_Fac is either 200 or the MinMax_Num, depending on the scale. For SFac < 1, it's the product of MinMax_Num and SFac.

6. **Base score and differentiation:**

- The **Base** score is calculated differently based on the **Sfac** value:

    - If **Sfac < 1**, the Base score is set to 150.

    - If 1 < Sfac < 100, the Base score directly uses the Sfac value.

    - If Sfac > 100, the Base score increases proportionally with every power of 10 (e.g., Base for Sfac of 500 is 105, for Sfac of 1000 is 110). This aims to reward high levels of service provision.

- An **UpdatedBase** is calculated only for **Sfac < 1** scenarios, introducing an additional fraction to differentiate scores in this range.

Base calculation: The Base is set to 150 for entities that have received more services than provided (SFac < 1). If the SFac is greater than 1 but less than or equal to 100, the Base is equivalent to SFac. For any SFac greater than 100, the Base is incrementally adjusted by adding fractions to the value of 100 to maintain proportionality and reflect the extent of services provided.

UpdatedBase calculation (UpdatedBase): UpdatedBase differentiates between the SFac values less than 1 and those greater than 100, adding a fraction based on the scale of SFac to the Base, which adds granularity and precision to the SES calculation for entities providing a higher volume of services.

7. **Finalizing the SES:**

- The **final Social Equity Score (SES)** is calculated by adding either the **UpdatedBase + Avg_MM_Fac** (if applicable) or the **Base + Avg_MM_Fac.**

In essence, the SES calculation considers various factors:

- Quantity and type of services provided and received

- Perceived social value of the service categories

- Individual circumstances and effort involved in providing services

Final SES calculation: The final SES is calculated by adding either the UpdatedBase (if it has a value) or the Base (if UpdatedBase is false, i.e., SFac is less than 1) to the Avg_MM_Fac. This results in the final SES, which is a comprehensive score reflecting the social equity contribution of an entity based on the services it provides and receives within its network.

A story behind how SES can be understood in real scenarios: Imagine a young man named Diego living in a vibrant neighborhood of São Paulo. Diego has always been passionate about community service, dedicating countless hours to helping local charities and organizing events to support underprivileged families. Despite his dedication, he struggles to find a stable job and often relies on the generosity of others to make ends meet.

Diego comes across a nonprofit initiative called "Helping Hands," which uses the Social Equity Score (SES) to connect individuals like him with various support services. Intrigued, Diego decides to participate, hoping to find a way to make a more significant impact while also receiving the help he needs.

Let's take this example from the data provided and explain the calculation of the SES. For clarity, we shall use the example from the category "Charity, Religious & Community Services."

Category: Charity, Religious & Community Services

Services provided (SP): Diego has provided 10 instances of community service, such as organizing food drives and volunteering at local shelters.

Services received (SR): In return, Diego has received support 20 times, including job training sessions and financial aid to cover basic needs.

Service category weightage (SW): 4 (which translates to 600 when multiplied by 150)

Service efforts (SE): 150

Service efforts rating: 90

Individual status rating: 50

Disability status rating: 75

After we run the secret sauce and apply AI/GenAI, the Social Equity Score (SES) for this example is ***321.25***, considering all the calculations based on the provided data and formulae.

With an SES of 321.25, Diego's efforts and the support he has received are quantified, highlighting his significant contributions to the community. This score not only reflects his dedication but also ensures he gains access to further support and opportunities. Through "Helping Hands," Diego is connected with additional resources, such as advanced job training and mentorship programs, which help him secure stable employment.

This story demonstrates the power of the SES in recognizing and rewarding individuals who contribute to society. By leveraging AI and Generative AI, "Helping Hands" ensures that individuals like Diego are supported and empowered, fostering a more equitable and supportive community.

This example showcases how the SES calculation incorporates multiple factors including the number of services provided and received, service category weightage, and updated service efforts. The formula ensures that the SES reflects both quantitative and qualitative aspects of social contributions, promoting a fair and balanced system.

In our example, the SES calculation integrates various elements to ensure that the final score represents a comprehensive view of the individual's social contributions and the support they receive. By leveraging AI and Generative AI, the system can refine the weighting of different variables, predict long-term impacts, and adapt to real-world complexities, ensuring a fair and equitable assessment.

By combining these factors, the SES aims to provide a **holistic assessment** of an individual or organization's contribution to social equity. It's important to remember that this is just a one specific example, and different methods and factors might be used in other contexts to calculate social equity scores.

This SES calculation algorithm, powered by AI, can process vast quantities of data and apply these steps accurately and efficiently. The AI component can identify patterns, suggest optimizations, and predict future SES outcomes based on changes in service patterns, thereby empowering entities to enhance their social equity impact proactively. AI-driven SES models can serve as crucial tools for organizations to align their service delivery and community engagement with broader social equity goals.

Again the steps, the variables, and the equations provided here are very basic to give an example, but a lot more logic goes into the final SES calculation which you can vouch for by reaching out to the author's startup mentioned in the last chapter!

How Do Recommendations Work with AI to Improve SES?

AI algorithms offer a sophisticated approach to analyzing and improving social equity scores, going far beyond simple data collection to offer actionable insights for organizations and individuals alike. By sifting through extensive datasets, these algorithms can pinpoint key areas where social responsibility can be improved, environmental impacts can be lessened, and ethical practices can be enhanced.

Consider an organization with an extensive carbon footprint due to reliance on nonrenewable energy sources. AI can evaluate the energy consumption patterns, correlate them with sustainable alternatives, and propose a comprehensive switch to greener energy solutions like solar or wind power. Such a move not only improves the organization's environmental impact but also boosts its social equity score, reflecting a commitment to planetary stewardship.

In socioeconomic terms, AI can address inequities by analyzing compensation structures within an organization. If the algorithm determines that employees are underpaid, it can prompt a reassessment of wage policies, potentially advocating for increases to ensure that all employees receive fair compensation. This adjustment not only aids the individuals directly affected by low wages but also signals to the community and stakeholders that the organization prioritizes fair labor practices.

Moreover, these AI systems can identify gaps in corporate social responsibility initiatives, suggesting new community engagement strategies or pinpointing opportunities for philanthropic endeavors that align with the company's values and the needs of the community.

The detailed recommendations provided by AI are rooted in robust data analysis, ensuring that suggestions are not just ethically sound but also feasible and effective. As organizations implement these AI-generated recommendations, they not only see improvements in their social equity scores but also contribute to a larger movement toward equity and justice, paving the way for a more ethical and socially responsible future in business and beyond.

Let's take an example to give the message: Imagine "Green Horizons," a landscaping company known for its beautiful designs, grappling with a dilemma. While their award-winning gardens flourish, their reliance on traditional gas-powered equipment leaves a significant carbon footprint. This not only impacts the environment but also raises concerns within the community about the company's social responsibility.

Enter "Empowerment Labs," a team of AI specialists pioneering the use of AI in promoting social equity. Utilizing advanced algorithms, they analyze "Green Horizons" data, examining things like fuel consumption, energy sources, and readily available alternatives. The results reveal a significant opportunity to improve.

Empowerment Labs presents "Green Horizons" with a comprehensive plan, powered by GenAI, that forecasts the impact of switching to electric equipment and incorporating solar panels. The data paints a clear picture: the switch, while requiring an initial investment, would not only significantly reduce their carbon footprint but also boost their social equity score by demonstrating their commitment to environmental sustainability. This aligns with the growing concerns of the community and strengthens "Green Horizons" reputation as a responsible business citizen.

However, the impact extends beyond environmental benefits. As "Green Horizons" transitions to cleaner energy sources, they discover unexpected cost savings. The reduced reliance on gas translates to lower fuel expenses, and the solar panels generate a portion of their energy needs, further reducing operational costs. Additionally, the company experiences a surge in employee morale, as staff takes pride in working for an organization committed to environmental responsibility.

This ripple effect extends further. "Green Horizons" becomes a role model for other businesses in the community, inspiring them to explore similar sustainable practices. They also partner with local schools, offering educational workshops on renewable energy sources and environmental stewardship.

This story exemplifies the power of AI in fostering social equity. By analyzing data and generating actionable insights, AI empowers organizations like "Green Horizons" to make informed decisions that benefit not only their business but also their employees, the environment, and the broader community. As AI and GenAI continue to evolve, they hold the potential to guide individuals and organizations on a path toward a more equitable and sustainable future, weaving together environmental responsibility, ethical practices, and a commitment to the betterment of society.

What Is the Tool or App That Uses This Secret Sauce?

In the intersection of technology and social justice, blockchain and AI stand as the twin pillars supporting innovative approaches to global equity challenges. Cryptocurrency, underpinned by blockchain's transparent and secure ledger, is no longer just a speculative asset but a vehicle for societal change, as evidenced by pioneering projects like GoodDollar. GoodDollar seeks to democratize wealth distribution through a universal basic income model, leveraging AI to ensure that the cryptocurrency reaches those who need it most. This fusion of blockchain and AI transcends traditional barriers to financial access, fostering a digital economy where opportunity and support are available to all, regardless of geographic or socioeconomic status.

Meanwhile, the Impact Lab integrates AI's analytical prowess with blockchain's integrity to holistically measure and enhance social equity. It navigates the complex terrain of social impact, offering a blockchain-verified trail of an organization's influence and AI-driven insights for sustainable improvement. This synergy between AI and blockchain extends beyond the realm of finance into every sector, advocating for a merit-based distribution of resources and recognition of impactful efforts.

The interplay of crypto, blockchain, AI, and GenAI is redefining what's possible in creating a socially equitable world. It's a world where the value of every action can be accurately measured, transparently reported, and equitably rewarded, anchoring the very essence of meritocracy in the digital age. These technologies together form a powerful toolkit for dismantling inequality and building a more inclusive future.

While initiatives like GoodDollar represent a commendable effort to democratize wealth distribution through blockchain and AI technologies, there remains a nuanced challenge in ensuring that the aspects of quality and merit are fully integrated into the framework of social equity. The pursuit of social equity is multifaceted, and while the distribution of resources is one critical component, it is not the only measure of a just and fair society.

To truly foster a balanced ecosystem, these technologies must not only facilitate access but also reward contributions that advance societal well-being and innovation. As we advance, it is paramount to refine these systems to recognize and incentivize quality contributions and merit-based achievements. This ensures that while we strive for equality in opportunity, we also preserve the integrity of effort and excellence, preventing a one-dimensional approach that may overlook the diverse aspects of individual and organizational value.

Hence, while blockchain and AI are potent tools for social change, their application must be continually assessed and calibrated. This ensures that the drive toward social equity does not inadvertently sideline the recognition of individual merit and the high standards of quality that propel society forward. It's about creating a balanced approach that values both equitable access and the exceptional contributions that individuals and organizations make to society.

XTrawConnect.com emerges as an innovative startup poised to revolutionize the way social equity is measured and enhanced in organizations and individuals alike. Leveraging the proprietary methodologies outlined in previous chapters and discussions, this platform is designed to operationalize the "secret sauce" that underpins the calculation of the Social Equity Score (SES). We will talk about this in the last chapter!

Conclusion

In wrapping up Chapter 7, we've delved deep into the "secret sauce" that artificial intelligence brings to the table in crafting and refining the Social Equity Score (SES). The intricate dance of algorithms and data paints a picture of potential – a world where equity is not just a vision but a quantifiable, achievable reality. This chapter stands as a testament to the power of AI in transforming abstract values into concrete, actionable insights that organizations and individuals can use to gauge and amplify their impact on society.

Peering ahead, the final chapter beckons us to consider the expansive horizon where language models elevate the SES from concept to cornerstone of a socially equitable world. It envisages a future where AI's linguistic prowess is harnessed to interpret, communicate, and evolve the narratives of social equity, reinforcing the virtuous cycle of giving rather than taking. This paradigm shift, powered by the most advanced language models, promises to weave the fabric of a society where every contribution is recognized and rewarded, and where every member is motivated to add value to the collective whole. As we approach the close of this explorative journey, the road ahead is luminous with the promise of innovation – a beacon guiding us toward a world rich in merit, equity, community, and shared success.

CHAPTER 8

The Way Ahead

As we venture into the final chapter of this exploration, we're on the brink of a new horizon where AI-driven language models can significantly bolster social equity. Envision a future where an app, powered by the most advanced AI and GenAI technologies, becomes a catalyst for change, meticulously designed to champion social equity. This chapter is dedicated to unraveling the potential of such an app, laying out a strategic roadmap for its development, and inviting you to be an integral part of this transformative journey.

The promise of AI in this context is not just in automating processes but in intelligently analyzing vast swathes of data to unearth insights that drive equitable outcomes. Picture a platform where every interaction, every transaction, is scrutinized through the lens of fairness, inclusivity, and merit. Here, AI isn't merely a tool; it's a collaborator that aligns with our highest ideals, ensuring that opportunities and resources are allocated justly, recognizing and rewarding the true value individuals and organizations bring to the table.

As we chart the course for this app, the roadmap is clear: develop AI frameworks that not only understand the nuances of human language but also grasp the complexities of social equity. The journey ahead is about refining these technologies to discern, with unprecedented precision, the contributions that advance societal well-being, ensuring that merit and quality are at the heart of the equity equation.

© Raghu Banda 2024
R. Banda, *Building Social Equity with AI*, https://doi.org/10.1007/979-8-8688-0091-7_8

Joining this movement isn't just about adopting new technology; it's about being part of a community committed to leveraging innovation for the greater good. As members and contributors, your insights, feedback, and active participation will be the cornerstone of this endeavor, shaping an AI ecosystem that's not only intelligent but also just and fair. Together, let's embark on this path, harnessing the boundless potential of AI to forge a society where equity is woven into the very fabric of our interactions and decisions.

How Can Language Models Play a Big Role in Social Equity?

Large language models (LLMs) like GPT-4 have the transformative potential to reshape our understanding and approach to social equity, embodying the nuanced perspectives of great AI thinkers of our times. These models, when adeptly integrated with AI and GenAI, can dissect and analyze vast arrays of data, extracting insights that are not immediately apparent and often overlooked in traditional analyses.

In a professional context, LLMs can revolutionize HR and recruitment processes, ensuring that opportunities are distributed based on merit and qualifications rather than biases or preconceived notions. By analyzing job descriptions, applications, and feedback, they can identify and mitigate language that may unconsciously favor certain demographics over others, fostering a more inclusive workplace that truly values merit and quality.

On a personal level, these models can offer individualized education and skill development pathways, acknowledging different learning styles and needs. For instance, they can provide tailored educational content that bridges gaps in understanding, allowing users from diverse backgrounds to achieve their full potential based on their abilities and efforts, not their starting points in life.

Societally, LLMs can contribute to policymaking by analyzing public opinions, debates, and discussions across various platforms to gauge the public's sentiment on social equity issues. They can help draft more equitable laws and policies by synthesizing a broad spectrum of voices and perspectives, ensuring that legislation is both fair and effective.

An example of this in action is the development of tools like AI-driven platforms that analyze legal documents to ensure they adhere to principles of equity and inclusivity, flagging potential biases. Similarly, AI can assist in urban planning, ensuring resources are allocated fairly across communities, analyzing various data points to recommend where schools, hospitals, and other essential services should be situated to serve all citizens equitably.

Imagine Sofia, a bright young woman from a disadvantaged background. Despite her intelligence and dedication, Sofia struggles to find opportunities that match her potential. Traditional job applications often focus on rigid criteria, failing to capture the nuances of her experience.

Enter "Equitable Pathways," a revolutionary platform powered by advanced LLMs. When Sofia uploads her resume, Equitable Pathways doesn't just scan for keywords. It delves deeper, analyzing her work history and achievements, identifying transferable skills, and highlighting her potential. The LLM, trained on vast amounts of data and nuanced perspectives on social equity, recognizes Sofia's grit and resourcefulness gained from overcoming challenges in her background.

Equitable Pathways goes beyond just analyzing Sofia's qualifications. It also analyzes job descriptions, identifying and mitigating unconscious biases present in the language. This ensures that opportunities are presented fairly based on merit, not demographics. As a result, Sofia receives interview invitations from companies that truly value her unique skillset and potential.

This impact isn't limited to Sofia. Equitable Pathways helps dismantle systemic barriers in workplaces, creating a more level playing field for everyone. Companies benefit from a more diverse talent pool, fostering innovation and creativity.

The influence of Equitable Pathways extends beyond the professional realm. The platform also offers personalized learning pathways. By analyzing Sofia's learning style and knowledge gaps, it recommends educational resources and programs tailored to her specific needs. This empowers Sofia to continuously develop her skills and bridge any gaps in her education, allowing her to reach her full potential.

On a broader societal level, Equitable Pathways anonymizes and aggregates data from its diverse users. This data empowers policymakers by providing insights into the challenges faced by different communities. By analyzing public sentiment and discussions on social media, the platform helps identify areas where policies need reform to be more equitable. Imagine Equitable Pathways informing the development of programs that address issues like unequal access to education or healthcare, ensuring resources are allocated fairly across communities.

The story of Equitable Pathways exemplifies the transformative power of LLMs. By going beyond mere numbers and delving into the complexities of human experiences, LLMs have the potential to become powerful tools for social equity. They can dismantle biases, create personalized opportunities, and inform fair and effective policies, paving the way for a future where everyone has the chance to thrive based on their merit and effort.

As we delve into the future, the integration of LLMs in our digital ecosystem offers a beacon of hope for enhancing social equity, underscoring the importance of merging human-centric values with cutting-edge technology to build a more just and equitable society.

Bridging Communication Gaps

Language models like GPT-4 are not just tools for processing text; they're bridges connecting diverse cultures and communities, embodying the analytical depth of Kara Swisher, the strategic foresight of the author, the practical insight of Steve Kunnon, and the visionary outlook of Elon Musk to name a few! These AI-driven systems offer a profound capacity to enhance mutual understanding and collaboration across the globe.

Professionally, these models can transform international business, enabling real-time translation and cultural nuance interpretation, thus breaking down the traditional barriers that have hindered global commerce and collaboration. A project manager in New York can instantly understand the concerns of a developer in Tokyo, ensuring that collaboration is based on mutual understanding and respect, driving toward shared goals with a sense of unity and purpose.

On a personal level, these language models can assist in everyday interactions, making learning new languages more accessible and fostering connections between individuals from different cultural backgrounds. They can serve as personal language tutors, adapting to the user's learning style, and providing instant feedback, thereby democratizing language learning and enabling more personal and meaningful cross-cultural interactions.

Societally, GPT-4 can aid in bridging divides by providing unbiased, balanced interpretations of different viewpoints, enabling a deeper understanding of various social issues. For instance, during a public health crisis, it can offer clear, accurate information and guidance to non-native speakers, ensuring that crucial information is accessible to all, irrespective of language proficiency.

One real-time example is the use of AI in emergency response systems, translating and interpreting distress calls from non-native speakers rapidly, ensuring that help is dispatched promptly and accurately. Another is the application in social media platforms, translating user-generated content in real time, allowing for a global exchange of ideas, stories, and support, thus nurturing a more inclusive digital community.

In this way, language models like GPT-4, Gemini, LLAMA, or Claude don't just translate words; they translate meanings, intentions, and emotions, fostering a world where quality communication is a universal right, not a privilege.

Enhancing Accessibility

In these realms of technology and social innovation, AI-driven language models like GPT-4 have emerged as powerful tools for enhancing accessibility and inclusivity. In the professional sphere, these models can transform workplaces by ensuring that all employees, regardless of visual or hearing impairments, have equal access to information and opportunities for collaboration. For example, audio descriptions and real-time transcriptions can facilitate more inclusive meetings and training sessions, allowing everyone to contribute to and benefit from collective knowledge and decision-making processes.

On a personal level, these advancements can profoundly impact individuals with disabilities by providing them with greater autonomy and access to digital content. Real-time transcriptions and audio descriptions of online courses, for instance, enable learners with diverse needs to pursue education and personal development on an equal footing with others, fostering a culture of continuous learning and self-improvement based on merit and interest rather than accessibility barriers.

Societally, the ripple effects of these technologies can be profound. For instance, museums and public institutions can use AI to offer guided tours with audio descriptions or sign language interpretation, making cultural and educational experiences more accessible to all community members. This not only democratizes access to knowledge and culture but also encourages a more inclusive and empathetic society where everyone's experiences and contributions are valued.

Such real-time AI applications underscore the potential of technology to bridge gaps in our society, ensuring that every individual, irrespective of their abilities, can partake in the benefits of our digital age. This vision aligns with the forward-thinking perspectives of the current generation innovators, who advocate for leveraging technology to create more equitable and inclusive environments across all facets of life.

Tackling Misinformation

In an era where information is as ubiquitous as it is potent, the role of AI-powered language models in ensuring the integrity and reliability of disseminated content is invaluable. These advanced tools can sift through vast datasets to identify and rectify inaccuracies, thereby playing a pivotal role in maintaining the factual accuracy that underpins a well-informed public discourse.

Professionally, this capability translates into more robust due diligence processes and enhanced accountability within industries. For instance, financial analysts can leverage AI to cross-verify market reports and investment insights, ensuring that decisions are made based on accurate and timely information, thus promoting merit-based outcomes and equity in financial success.

On a personal level, individuals benefit from AI's fact-checking prowess by being better equipped to make informed decisions, from healthcare choices to political opinions. For example, an individual researching health information online would receive verified content, reducing the risk of misinformation influencing personal or family health decisions.

Societally, the impact is profound. In the political arena, AI tools can analyze speeches, campaign messages, and news articles to identify and highlight misinformation, empowering citizens to base their voting decisions on verified facts. This fosters a more equitable democratic process where public opinion and policy are shaped by truth rather than fabrications.

A real-time example of this in action is the use of AI in monitoring social media platforms to flag and correct false information. Such applications not only protect users from being misled but also uphold the quality and merit of public discourse, contributing to a more equitable and informed society. In this context, the perspectives of the current tech founders and stalwarts converge on the potential of AI to serve as a guardian of truth, reinforcing the foundations upon which a fair and just society is built.

Personalized Education

Harnessing the potential of AI-powered language models, we can revolutionize the educational landscape, making it more inclusive and adaptive. These models offer the capability to tailor educational content to meet the unique needs of each learner, acknowledging their pace, style, and level of understanding. In professional settings, this translates to more effective training programs where employees, regardless of their prior knowledge or background, receive customized learning experiences that enhance their skills and productivity.

On a personal level, imagine a student struggling with a particular concept in mathematics. An AI-driven language model can not only identify the student's specific challenge but also curate or generate targeted exercises and explanations, providing instant feedback that guides the student toward mastery. This individualized support can bridge learning gaps, making quality education accessible to all, irrespective of their socioeconomic status.

Societally, the implications are profound. In regions where educational resources are scarce, AI-powered language models can deliver high-quality educational content, breaking down the barriers imposed by geographic and economic constraints. For example, a language model can provide real-time language translation and tutoring to students in remote areas, offering them the same quality of language education as in urban centers.

A real-time example of this transformative potential is seen in platforms that use AI to offer adaptive learning paths, such as Khan Academy or Coursera, which adjust their content based on user interaction and performance. By providing personalized learning experiences, these platforms ensure that every learner, regardless of background, has the opportunity to excel and develop their potential.

In essence, the perspectives of our generation forward tech thinkers converge on the belief that AI-driven language models are pivotal in democratizing education, ensuring that quality and merit define one's educational journey, rather than their starting point in life.

Social Equity App with AI and the Roadmap

XTrawconnect.com emerges as an innovative startup poised to revolutionize the way social equity is measured and enhanced in organizations and individuals alike. Leveraging the proprietary methodologies outlined in previous discussions, this platform is designed to operationalize the "secret sauce" that underpins the calculation of the Social Equity Score (SES).

This platform encapsulates a comprehensive framework for achieving SES by harnessing the intricate algorithms and multifaceted variables detailed in the last chapter. Its core functionality is built upon the thorough analysis of service contributions, both provided and received, and the weightage of various service categories. XTrawconnect.com's sophisticated

AI-driven system meticulously calculates the SES by integrating diverse factors such as environmental stewardship, ethical governance, and social responsibility into a unified score.

At the heart of XTrawconnect.com's vision is the goal to create a transparent and equitable metric that not only quantifies social contributions but also drives organizations toward meaningful action. By providing clear insights and actionable recommendations, XTrawconnect.com aims to guide its users on a path to enhanced social impact and a more sustainable future. Whether for corporate entities or individuals seeking to quantify and improve their societal footprint, XTrawconnect.com stands as a beacon of innovation in the pursuit of social equity.

To make this more compelling, let's take a hypothetical example which might be a reality in future: Imagine Dr. Anya Chandra, a passionate medical professional dedicated to improving healthcare access in underserved communities. While she volunteers at a local clinic and provides pro bono services, there's always a lingering question: "Is what I'm doing truly making a difference?"

Enter XTrawconnect.com, a revolutionary platform leveraging AI to measure and empower social equity. Intrigued, Dr. Chandra signs up and begins logging her volunteer hours, educational outreach programs, and collaborations with local NGOs.

XTrawconnect.com's AI analyzes her contributions, considering not just the quantity of her actions but also the quality and impact. It factors in the difficulty of outreach programs in certain communities, the specific needs of the population served, and even the potential long-term impact of her initiatives.

Dr. Chandra receives a personalized Social Equity Score (SES), reflecting her dedication to improving healthcare access. But XTrawconnect.com doesn't stop there. It analyzes Dr. Chandra's location, identifies nearby resource gaps, and suggests new avenues for impactful contributions. It recommends partnering with a mobile health clinic to reach remote communities, suggesting relevant training programs to enhance her skills, and even connecting her with other individuals passionate about healthcare equity.

Empowered by these insights, Dr. Chandra joins forces with other healthcare professionals identified by XTrawconnect.com and launches a mobile clinic initiative. Together, they provide crucial medical services to previously underserved areas, directly impacting the lives of countless individuals.

The impact extends beyond Dr. Chandra and the communities she serves. XTrawconnect.com anonymizes and aggregates data from its diverse users, providing valuable insights to policymakers and organizations. This data helps identify systemic inequalities in resource allocation, informs the development of targeted social programs, and sheds light on the most effective approaches to community development.

XTrawconnect.com becomes a catalyst for a collective movement toward social equity. By valuing quality alongside quantity, it fosters meaningful action, empowers individuals like Dr. Chandra to maximize their impact, and guides organizations and policymakers toward creating a more equitable future for all. This exemplifies the potential of AI and GenAI to rise above mere numbers and truly empower meaningful social change.

How Can You Become a Member?

Becoming a member of XTrawconnect.com offers you the unique opportunity to be part of a pioneering community dedicated to fostering social equity through the innovative application of technology. By joining, you're not just enhancing your own Social Equity Score (SES); you're contributing to a larger movement that values fairness, inclusivity, and merit in every interaction. Whether you're an individual looking to make a positive impact or an organization aiming to enhance your social responsibility, XTrawconnect.com provides a platform to track, improve, and leverage your contributions toward creating a more equitable society.

XTrawconnect.com is poised to revolutionize how we perceive and contribute to social equity through its innovative SES (Social Equity Score) system. By becoming a member, you can actively engage in shaping a more equitable future while tracking and enhancing your social impact. Here's a breakdown of the membership tiers available, designed to cater to a diverse range of needs and preferences:

Free membership: Ideal for newcomers to the concept of social equity scoring, as a free member, you can

- Record the services you provide and receive, creating a foundational understanding of your Social Equity Score

- Access basic insights into your SES, offering a glimpse into how your interactions contribute to broader social equity

- Engage with a community committed to social impact, sharing experiences and learning from others

Basic membership: For a minimal fee, the basic membership offers a deeper insight into the SES. Along with all the benefits of the free membership, you gain

- Detailed explanations of your SES, understanding the nuances of how your score is calculated and the factors influencing it

- Access to educational content designed to enhance your awareness and ability to contribute positively to social equity

- Firsthand insights into the methodology behind SES, fostering a more profound understanding of its significance

Premium paid membership: The premium tier is the most comprehensive package, offering advanced features and in-depth analysis. In addition to recording and understanding your SES, this level predicts future scores based on various service scenarios and recommendations. The premium membership provides comprehensive features:

- Predictive insights into how different actions might influence your future SES, enabling strategic planning to enhance your social equity contribution

- Personalized recommendations and strategies based on sophisticated analysis of various service weightages and transaction categories

- Exclusive access to advanced tools and features designed to maximize your impact and understanding of social equity dynamics

- Priority support and access to community experts, along with opportunities to influence future enhancements of XTrawconnect.com

Each tier is designed to accommodate different levels of engagement and interest in building social equity, allowing you to choose the path that best aligns with your aspirations and commitment. Whether you're starting with the free tier to explore the basics or diving deep with the premium membership to maximize your impact, XTrawconnect.com offers a structured and insightful journey toward understanding and improving your Social Equity Score.

By choosing to become a member of XTrawconnect.com, you are stepping into a future where technology empowers us to quantify and improve our social contributions, driving a more equitable and meritocratic world. Whether you start with a free membership to get your feet wet or dive into the advanced features of our premium service, your journey toward building a more equitable society begins here.

Conclusion

As we conclude Chapter 8, we look forward with optimism to the potential that the SES app, augmented by large language models, holds for the future of social equity. This innovative tool, underpinned by advanced AI and blockchain technology, promises a new horizon where measuring, understanding, and enhancing social equity becomes an integrated part of our digital experience. The SES app is not just a technological advancement; it is a step toward a future where every individual and organization can actively track and improve their contribution to a fairer world.

Reflecting on the entirety of this book, we've embarked on a comprehensive journey through the landscape of social equity in the age of technology. From the initial discussions on the importance of social equity and the role technology can play to the deep dives into the Social Equity Score (SES) and its implications across various domains, the book provides a thorough exploration of how we can leverage technology to foster a more equitable society.

The overarching narrative weaves together the threads of AI, blockchain, and large language models, illustrating their transformative potential in enhancing our understanding and practice of social equity. By demystifying these technologies and showcasing their application in building the SES, the book empowers readers to envision and work toward a future where technology serves as a catalyst for social good.

As we close this chapter and the book, the journey doesn't end here. The insights gained are but stepping stones toward active participation in shaping a world where equity, quality, and merit are intertwined in the very fabric of our social and professional interactions. This book invites readers to not only conceptualize but also actively engage in creating a future where technology amplifies our collective ability to build a just, inclusive, and equitable society.

In this narrative of progress and possibility, each reader is encouraged to take the concepts, ideas, and strategies discussed and apply them in their own spheres of influence. As technology continues to evolve, so too should our approaches to fostering social equity, always with an eye toward creating a world that values and uplifts every individual. Through this book, we gain not just knowledge but a call to action, to be architects of a future where technology and humanity converge in the pursuit of a more equitable world.

Index

A

Academic exploration, 42
Academic research, Gephi, 159
Adaptive AI algorithms, 87
Adaptive capacity, 87, 88
Adaptive learning, 102
Affordable housing programs, 6, 10
AI-driven alumni networks, 145
AI-driven collaboration tools, 136
AI-driven diagnostic tools, 121
AI-driven features, 75
AI-driven network connections
  AI matchmaking tool, 132
  professional development
    networks, 132
  virtual networking platform, 132
AI-driven network expansion
  strategies, 140, 141
AI-driven networking app, 141
AI-driven networking initiatives,
  127, 128
AI-driven online education
  platform, 121
AI-driven professional
  platforms, 153
AI-driven support initiatives within
  networks

AI-enhanced peer review, 137
  diverse talent recruitment, 136
  personalized learning
    experiences, 137
  real-time translation, 137
  workload distribution, 138
AI-driven tools, 141
AI-enhanced communication
  platforms, 136
AI-enhanced networking, 146
AI-enhanced networking and
  alumni tools
  alumni networking
    platforms, 147
  inclusive professional
    networking, 148
  mentorship matching
    platforms, 148
  talent discovering
    platforms, 147
AI/GenAI, 103, 104, 106, 109
AI in vaccine distribution, 93
AI-powered impact assessment
  tool, 188
AI-powered language models, 235
AI-powered matchmaking, 131
AI-powered networking app, 143

I, J, K

L